Matrix Analysis of Structures

Matrix Analysis of Structures

Robert E. Sennett
California Polytechnic State University, San Luis Obispo

Long Grove, Illinois

For information about this book, contact:
Waveland Press, Inc.
4180 IL Route 83, Suite 101
Long Grove, IL 60047-9580
(847) 634-0081
info@waveland.com
www.waveland.com

Reissued 2000 by Waveland Press, Inc.

10-digit ISBN 1-57766-143-5
13-digit ISBN 978-1-57766-143-6

Printed in the United States of America

10 9 8 7

CONTENTS

PREFACE

Matrix analysis of structures and the closely related finite element method have achieved wide acceptance and use in virtually all engineering disciplines. The use of these methods, which require the solution of large numbers of simultaneous equations, has become a practicality due to the development of digital computers. Ever since the introduction of desktop microcomputers, these methods have been used in most engineering offices to solve a wide variety of problems.

The purpose of *Matrix Analysis of Structures* is to present to the student the displacement method of matrix analysis. Background required consists of basic strength of materials and introductory courses in structural analysis. The text should therefore be suitable for senior-level undergraduates, beginning graduate students, and practicing engineers in disciplines including civil, mechanical, architectural, and agricultural engineering.

The text begins with the development of the matrix method for the one-dimensional bar element. Because the bar or rod element is very simple, it allows development of all the necessary procedures without involving topics such as coordinate transformations. These transformations are derived when required for the element being considered. Thus, two-dimensional coordinate transformations are introduced in Chapter 2 for the truss and in Chapter 3 for the frame. Three-dimensional transformations are discussed in Chapter 5 (trusses) and in Chapter 6 (frames).

The distinction between matrix analysis of structures and the finite element method applied to structural problems is somewhat artificial since the same basic concepts and procedures are used in both cases. The primary difference between these methods lies in the procedures used for deriving the stiffness of the individual elements; that is, the determination of the force-displacement relationships for a specific type of element such as a bar, beam, plate, or shell element.

In the matrix analysis method, these relationships are generally determined by using strength of materials and basic structural theory. Thus, the method is limited to simple shapes such as rod, beam, and frame elements. In contrast, the finite element method is thought of as using more complicated elemental shapes such as plate or shell elements. The procedures used for deriving the force-displacement relationships for these elements generally involve minimizing a functional such as total energy.

Of course, these minimization procedures can also be applied to the simpler elements. This text uses, in addition to the structural analysis approach, the principles of virtual work and minimum potential energy to derive the elemental stiffnesses and global stiffness equations. In addition, topics such as non-nodal forces and the assemblage of elements can be developed easily using these techniques.

The author believes that computer programming is an essential part of learning the material presented. Naturally, the student is not expected to write commercial-quality code; however, the organization of program steps and understanding of algorithms is greatly enhanced by a hands-on programming problem. As a result, code fragments permeate the text in areas where procedures and algorithms are discussed, and problems for most chapters include the writing and use of a computer program for analyzing structures using the type of element presented in that chapter. And, once one program has been written, it can be easily modified and extended to deal with more complicated elements by simply addressing additional elemental degrees of freedom, coordinate transformations, and the differences in the elemental stiffness matrices. In this way, the author hopes that the student will gain a firm grasp of the techniques used in the ever-increasing number of commercial structural analysis and design computer programs available.

Although special topics and energy methods are presented in Chapters 7 and 8, respectively, some instructors may want to include sections of these chapters at earlier stages of development of the method. Thus, derivation of elemental stiffnesses and treatment of non-nodal loads using the principle of virtual work could be introduced in parallel with the development of material presented in Chapters 1 and 3. In addition, material presented in Chapter 7 such as elastic and inclined supports, and hinges in beam and frame members, could be considered when discussing the truss, beam, and frame chapters.

Finally, Chapter 9 presents a brief introduction to the finite element method. This is accomplished by using work and energy methods to derive the stiffness matrix and structural stiffness equation for the three-node triangular element.

The author is indebted to a number of reviewers for their useful comments and valuable suggestions. Specifically, the author wishes to thank Dr. Jack H. Emanuel, University of Missouri, Rolla; Dr. William J. Hall, University of Illinois; Dr. Ronald S. Harichandran, Michigan State University; Dr. Eric Lui, Syracuse University; and Dr. Everett McEwen, University of Rhode Island.

Robert Sennett

CHAPTER 1

ANALYSIS OF ONE-DIMENSIONAL BARS

1.1 INTRODUCTION

In matrix analysis of structures, two formulations are possible:

(1) the force method (flexibility method)

(2) the displacement method (stiffness method)

The difference between these formulations involves the selection of the variables used as unknown quantities. The force method, as you might expect, uses forces as the unknowns for which a solution is desired. Similarly, displacements are the unknowns in the displacement method of analysis. We can identify the classical methods of structural analysis with one or the other of these techniques. For example, when using the force method the equilibrium equations are first used. Additional equations are found by introducing compatibility conditions. Consider the beam shown in Figure 1-1.

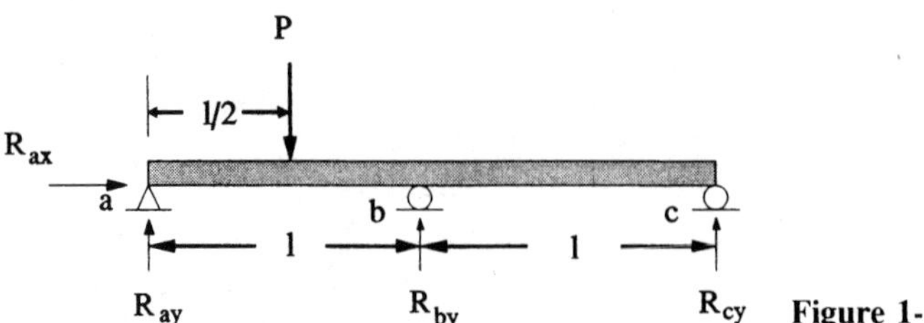

Figure 1-1

The beam shown is statically indeterminate to the first degree. That is, there are four possible reactions and only three equilibrium equations available. The possible reactions present at any support can always be determined by determining the displacements that are restrained. A reaction is possible in a given direction if the displacement in that direction is restrained. For the beam shown, the horizontal displacement is prevented at point a and the vertical displacements are prevented at all three support points. Thus, four reactions are possible. Since we are considering a two-dimensional problem, only three independent equilibrium equations are available. If we write the equilibrium equations we find:

$$\sum F_x = 0 \qquad R_{ax} = 0$$

$$\sum F_y = 0 \qquad R_{ay} + R_{by} + R_{cy} = 0.$$

$$\sum M_a = 0 \qquad R_{cy}(2l) + R_{by}(l) - P(l/2) = 0$$

Notice that the last two equations contain three unknowns. Clearly, we need an additional equation.

One classical technique for finding a solution to this problem is the method of *superposition.* As you will recall from your structural analysis coursework, the first step in applying this method is to select a redundant (or redundants if we have a multi-degree of indeterminacy problem) that, when removed, results in a stable, determinate structure. This is called our "primary" structure. Suppose we select the vertical reaction at point b as our redundant. We next draw a superposition diagram as shown in Figure 1-2.

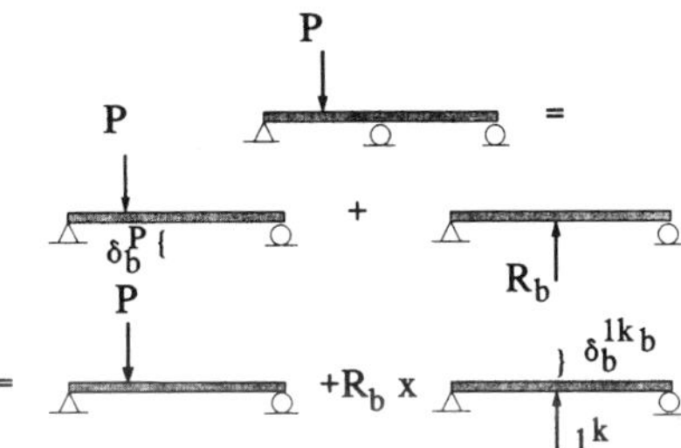

Figure 1-2 Superposition diagram.

From this diagram we can see that in order for the superposition to be valid we must insist that the total vertical deflection at point b is zero in order to match the zero displacement at this point on the original structure. Thus, $\delta_b^P + R_b \; x \; \delta_b^{1kb} = 0$. This is a compatibility of displacement equation that, when solved simultaneously with the two remaining equilibrium equations, yields the three unknown forces. Thus, the method of superposition is an example of the force method of analysis.

This problem can also be solved by other techniques such as *slope-deflection* or a technique derived from it, *moment distribution.* Again recall from your structural analysis courses that when using the slope-deflection method the unknowns are the joint rotations and translations. Equilibrium of the joints yields equations in these unknown displacements. Having found the displacements, the slope-deflection equations are then used to determine the forces (moments) at the joints, or nodes, of the members. Thus, the method of slope-deflection is an example of the displacement approach to solving structural problems.

The displacement method deals with kinematic indeterminacy rather than static indeterminacy. The degree of static indeterminacy of a structure is equal to the number of equations, in addition to the equilibrium equations, necessary to determine completely the reaction and member forces.

The degree of kinematic indeterminacy is equal to the total number of degrees of freedom of a structure that must be constrained to ensure zero displacements of the structure, excluding boundary constraints.

As an example, consider the two structures in the figure that follows.

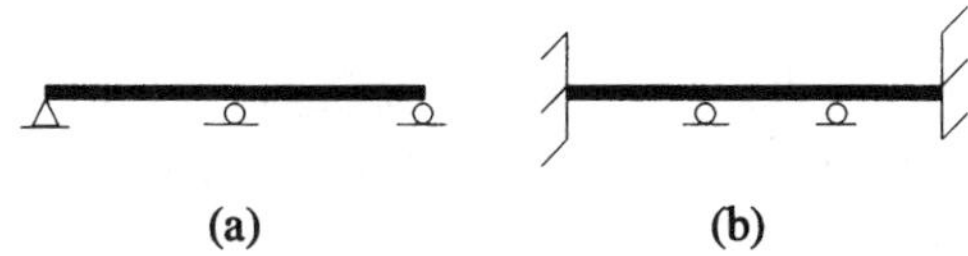

(a) (b)

If axial deformation is neglected, which is generally the case for beams, (a) is one-degree statically indeterminate (4 reactions, 3 equilibrium equations), and three degrees kinematically indeterminate (3 rotations at the supports). Part (b) has four degrees of static indeterminacy (4 vertical reactions and 2 moments at the fixed ends), but only two degrees of kinematic indeterminacy (2 rotations at the intermediate supports).

The formulation of a structural problem is generally much simpler when the displacement (or stiffness) method is used. Note that the more highly statically indeterminate the structure, the fewer displacements need to be found. Of course, the number of displacements that need to be found represents the number of simultaneous equations that must be solved. Most commercial computer programs use the stiffness approach, and we will use only this method in this text.

1.2 NODES AND DEGREES OF FREEDOM

As we have seen, the stiffness method uses displacements as the unknown quantities. Furthermore, these displacements are those of the nodes of the members. The nodes are points at which equilibrium will be enforced and displacements found. They are generally located at the ends of the members for most common structural shapes such as rods or beams. However, formulations are possible that place nodes at interior points of a member. We have noted in the previous section that the number of degrees of freedom at a node is equal to the number of possible displacements of that node. That is, the number of displacements that need to be specified in order to define uniquely the position of a node equals the number of degrees of freedom of that node. Therefore, for a one-dimensional bar or rod element, where displacements are restricted to a translation in the axial direction of the bar, each node has a single degree of freedom. For a two-dimensional truss element, each node has two degrees of freedom that correspond to translations in two directions (say x and y). For a two-dimensional beam element, we have one translation and one rotation possible at each node, yielding two degrees of freedom per node. For a two-dimensional frame element, each node has the capability of translating in two directions and rotating about one axis. Thus we have three degrees of freedom for each node. Since displacements constitute the unknowns in the stiffness formulation of a problem, the number of degrees of freedom will be indicative of the number of simultaneous equations that we will eventually need to solve. Of course, each structure has to have a number of restraints (such as supports) in order to remove any rigid body motion that could be present. The actual number of equations that must be solved is equal to the total number of degrees of freedom of the structure minus the number of restraints present.

1.3 STIFFNESS DEFINITION

If we have a simple linear spring fixed at one end with a force F applied at the other end, we know that the relationship between the applied force and the resulting displacement is $F = kx$ where x is the displacement and k is called the *spring constant* or the *spring stiffness*. We have a linear spring since the force-displacement relationship is linear with x. Notice that k is the force corresponding to a unit displacement $x = 1$. Of course, we can expand the concept of force-displacement relationships to structures with multiple degrees of freedom. When we do this, we generally write these relationships in matrix form, $\{F\} = [k]\{u\}$. For a two-degree-of-freedom system this matrix equation is expanded as shown below:

$$\begin{Bmatrix} F_1 \\ F_2 \end{Bmatrix} = \begin{bmatrix} k_{11} & k_{12} \\ k_{21} & k_{22} \end{bmatrix} \begin{Bmatrix} u_1 \\ u_2 \end{Bmatrix}$$

where the k_{ij}'s are called the *stiffness influence coefficients* and form the stiffness matrix.

Suppose that we let $u_1 = 1$ and $u_2 = 0$. Then $F_1 = k_{11}$. That is, k_{11} is the force at point 1 corresponding to a unit displacement at point 1 and only point 1. Similarly, k_{21} is the force at point 2 corresponding to a unit displacement at point 1 only. In fact, in general, the stiffness element k_{ij} is defined as the force at i corresponding to a unit displacement at j and j alone. This force is that required to maintain equilibrium in the displaced configuration. We will now use this definition to determine the stiffness matrix for a simple structure. Consider the following problem:

The structure shown in Figure 1-3 consists of two bar elements with individual stiffnesses (spring constants) k_1 and k_2, and three nodes. The two elements are connected at node 2. We wish to determine the stiffness matrix for the entire structure using the definition of stiffness. Notice that the structure has three degrees of freedom; three translations, one at each node in the axial direction of the bars. Thus, we will generate a 3×3 stiffness matrix.

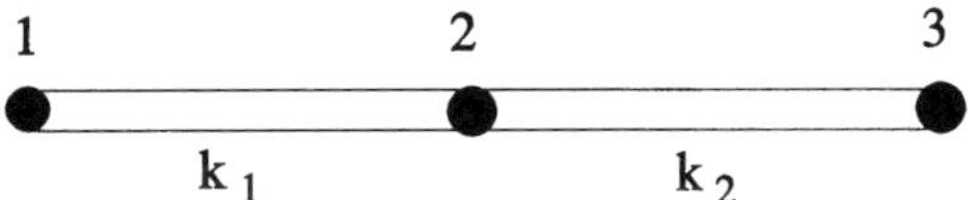

Figure 1-3 Two-element structure.

To accomplish our task it is necessary to introduce a unit displacement at each node, one at a time, and find the forces that must be present at the nodes in order to maintain equilibrium. Forces and displacements will be considered positive when acting toward the right.

Introduce a unit displacement at node 1, keeping all other nodes at their original positions. Figure 1-4 shows the free-body diagrams of the nodes and elements for this configuration.

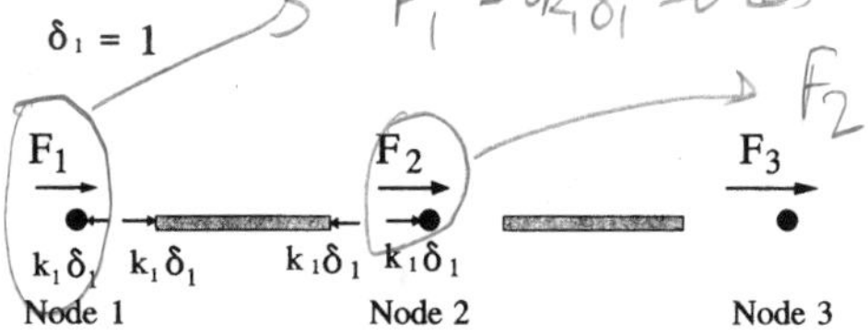

Figure 1-4 Unit displacement at node 1.

To maintain equilibrium of node 1, force F_1, which is the force at node 1 corresponding to a unit displacement at node 1, or k_{11}, must be equal to $k_1\delta_1$. Thus,

$k_{11} = k_1\delta_1 = k_1$ (since $\delta_1 = 1$). For equilibrium of node 2, force F_2, which equals k_{21}, must be equal to $-k_1\delta_1$. Thus $k_{21} = -k_1\delta_1 = -k_1$.

Since there are no displacements of nodes 2 and 3, there is no force in element 2 and F_3, which equals k_{31}, is zero. Thus, $k_{31} = 0$.

We next introduce a unit displacement at node 2. The corresponding free-body diagrams are shown in Figure 1-5.

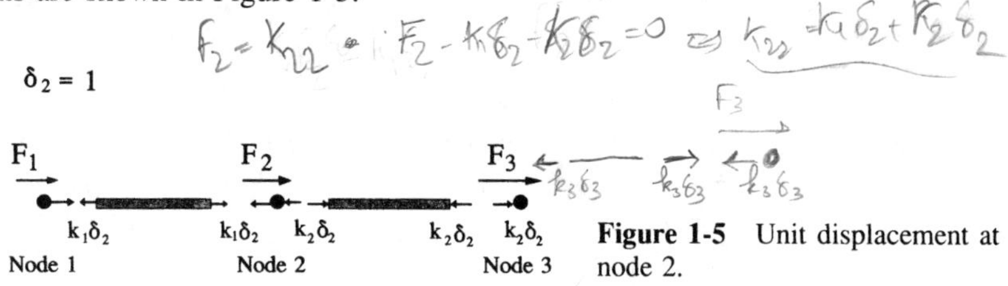

Figure 1-5 Unit displacement at node 2.

Proceeding in a manner similar to the previous case, realizing that $\delta_2 = 1$, we find

$$F_1 = k_{12} = -k_1, \quad F_2 = k_{22} = k_1 + k_2, \quad \text{and } F_3 = k_{32} = -k_2.$$

Finally, we introduce a unit displacement at node 3. Figure 1-6 shows the free body diagrams corresponding to this case.

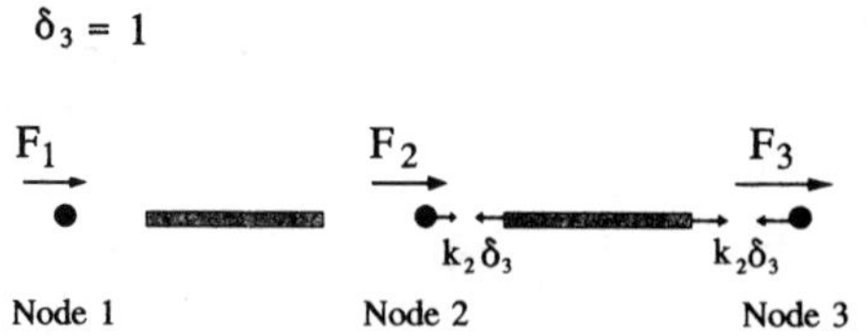

$\delta_3 = 1$

F1 F2 F3

$k_2\delta_3$ $k_2\delta_3$

Node 1 Node 2 Node 3

Figure 1-6 Unit displacement at node 3.

We find

$$F_1 = k_{13} = 0, \quad F_2 = k_{23} = -k_2, \quad \text{and } F_3 = k_{33} = k_2.$$

Thus the 3×3 stiffness matrix for this structure is

$$[K] = \begin{bmatrix} k_1 & -k_1 & 0 \\ -k_1 & k_1 + k_2 & -k_2 \\ 0 & -k_2 & k_2 \end{bmatrix}$$

We can achieve the same result by superposition of the forces in Figures 1-4, 1-5, and 1-6 when arbitrary displacements δ_i are imposed.

To illustrate, from Figure 1-4 where a displacement δ_1 is introduced, we have

$$F_1 = k_1\delta_1, \quad F_2 = -k_1\delta_1, \quad \text{and } F_3 = 0.$$

For a displacement δ_2, Figure 1-5 yields

$$F_1 = -k_1\delta_2, \quad F_2 = (k_1 + k_2)\delta_2, \quad \text{and } F_3 = -k_2\delta_2.$$

Finally, for a displacement δ_3, Figure 1-6 gives

$$F_1 = 0, \; F_2 = -k_2\delta_3, \quad \text{and } F_3 = k_2\delta_3.$$

Since we are dealing with a linear system, superposition is valid. Thus, if we introduce all three displacements simultaneously, the nodal forces required to maintain equilibrium will be the sum of the forces obtained by introduction of each displacement separately.

Adding the forces yields

$$F_1 = k_1\delta_1 - k_1\delta_2 + (0)\delta_3$$

$$F_2 = -k_1\delta_1 + (k_1 + k_2)\delta_2 - k_2\delta_3$$

$$F_3 = (0)\delta_1 - k_2\delta_2 + k_2\delta_3$$

Writing the above equations in matrix form yields

$$\begin{Bmatrix} F_1 \\ F_2 \\ F_3 \end{Bmatrix} = \begin{bmatrix} k_1 & -k_1 & 0 \\ -k_1 & k_1+k_2 & -k_2 \\ 0 & -k_2 & k_2 \end{bmatrix} \begin{Bmatrix} \delta_1 \\ \delta_2 \\ \delta_3 \end{Bmatrix}$$

or

$$\{F\} = [K]\{\delta\}.$$

Suppose we attempt to solve this equation for the displacements δ_1 through δ_3.

$$\{\delta\} = [K]^{-1}\{F\}$$

Clearly, we need to calculate the inverse of the structural stiffness matrix $[K]$. In order for a matrix to possess an inverse, its determinant must be non-zero (see Appendix A). Calculating the determinant of $[K]$ by using Cramer's rule and expanding using the first row we have

$$|K| = k_1 \begin{vmatrix} k_1 + k_2 & -k_2 \\ -k_2 & k_2 \end{vmatrix} - (-k_1) \begin{vmatrix} -k_1 & -k_2 \\ 0 & k_2 \end{vmatrix}$$

$$= k_1(k_1k_2 + k_2^2 - k_2^2) + k_1(-k_1k_2) = 0$$

Thus, the stiffness matrix is singular. This means that we have an infinite number of solutions to the equation. Referring to Figure 1-3, we note that there have not been any constraints imposed on the structure. Physically, with no support conditions specified, rigid body motion can occur, resulting in an infinite number of possible displaced configurations. As we shall see, once support conditions are introduced, a unique solution will be possible as long as the structure is a stable one.

Note that the stiffness matrix is symmetrical and has positive terms on the main diagonal. This will always be the case in structural problems. Positive terms must be on the main diagonal since a positive displacement at a node requires a positive force at that node. Also note that the terms in each column add to zero. This is a direct consequence of nodal equilibrium for unit displacements of each node individually.

1.4 INDIVIDUAL ELEMENT STIFFNESS

In the previous section we derived the stiffness matrix for a two-member structure directly. Since our ultimate goal is to apply the matrix analysis approach to systems with large numbers of degrees of freedom, we would like to generate the structural stiffness matrix for any structure by combining stiffnesses of individual members. Toward this goal we now derive the force-displacement relationships for a single bar element. That is, $\{F\} = [k]\{u\}$.

Consider the single element shown in Figure 1-7.

Figure 1-7 Single-bar element.

Clearly, this element has two nodes and therefore two degrees of freedom. We need to determine the elements of the stiffness matrix $[k]$ in the equation

$$\begin{Bmatrix} F_L \\ F_R \end{Bmatrix} = \begin{bmatrix} k_{11} & k_{12} \\ k_{21} & k_{22} \end{bmatrix} \begin{Bmatrix} u_L \\ u_R \end{Bmatrix}$$

Note that when $u_R = 0$ and $u_L = 1$, $F_L = k_{11}$ and $F_R = k_{21}$, which agrees with our definition of stiffness k_{ij} as the force at i corresponding to a unit displacement at j. In addition, when $u_R = 1$ and $u_L = 0$, $F_L = k_{12}$ and $F_R = k_{22}$. Therefore, to find the elemental stiffness terms, we need to introduce unit displacements at each end of the bar, one at a time, and find the corresponding forces.

Recall from basic mechanics of materials that the change in length of a prismatic bar loaded by an axial force F at its end is given by the expression $\delta = FL/EA$. Introducing a unit displacement at only the left end yields $u_L = F_L L/EA$. Thus, $F_L = EA/L = k_{11}$. Also, $F_R = -F_L = -EA/L = k_{21}$.

Next we introduce a unit displacement at only the right end giving $u_R = 1 = F_R L/EA$. Thus, $F_R = EA/L = k_{22}$. $F_L = -F_R = -EA/L = k_{12}$. The elemental stiffness matrix becomes

$$[k] = \begin{bmatrix} EA/L & -EA/L \\ -EA/L & EA/L \end{bmatrix} = EA/L \begin{bmatrix} 1 & -1 \\ -1 & 1 \end{bmatrix}$$

and the force-displacement relationship for this one-dimensional bar or rod element becomes

$$\begin{Bmatrix} F_L \\ F_R \end{Bmatrix} = EA/L \begin{bmatrix} 1 & -1 \\ -1 & 1 \end{bmatrix} \begin{Bmatrix} u_L \\ u_R \end{Bmatrix}$$

Note that in the force-displacement relationship it is important to keep the order of forces and displacements as left first, and then right, since this is the order for which the elemental stiffness matrix was derived.

1.5 COMBINATION OF ELEMENT STIFFNESSES

Figure 1-8 shows free-body diagrams of the members and joints of the two-member structure shown in Figure 1-3.

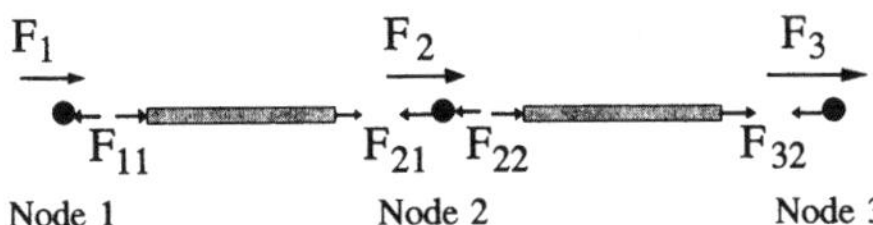

Figure 1-8 Free-body diagrams of elements and nodes.

Note that the F_i forces are either applied nodal forces or reactions. From joint equilibrium, $F_1 = F_{11}$, $F_2 = F_{21} + F_{22}$, and $F_3 = F_{32}$. Also note that the forces F_{ij} correspond to forces at the ith node of the jth element. Since we cannot specify both a force and a displacement at the same node, at each node we will know either the applied nodal force or the nodal displacement. To illustrate the combination of elemental stiffnesses, consider the example shown in Figure 1-9.

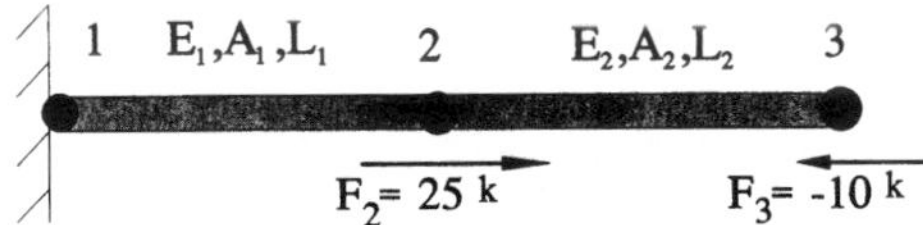

Figure 1-9 Two-element bar structure.

Example 1.1

Referring to Figure 1-8 and using the elemental force-displacement relationships we have:

Element 1:

$$\begin{Bmatrix} F_{11} \\ F_{21} \end{Bmatrix} = (E_1A_1/L_1)\begin{bmatrix} 1 & -1 \\ -1 & 1 \end{bmatrix}\begin{Bmatrix} u_1 \\ u_2 \end{Bmatrix}$$

Element 2:

$$\begin{Bmatrix} F_{22} \\ F_{32} \end{Bmatrix} = (E_2A_2/L_2)\begin{bmatrix} 1 & -1 \\ -1 & 1 \end{bmatrix}\begin{Bmatrix} u_2 \\ u_3 \end{Bmatrix}$$

Letting $E_iA_i/L_i = k_i$ we have

$$\begin{Bmatrix} F_{11} \\ F_{21} \end{Bmatrix} = k_1\begin{bmatrix} 1 & -1 \\ -1 & 1 \end{bmatrix}\begin{Bmatrix} u_1 \\ u_2 \end{Bmatrix} \tag{1.1}$$

$$\begin{Bmatrix} F_{22} \\ F_{32} \end{Bmatrix} = k_2\begin{bmatrix} 1 & -1 \\ -1 & 1 \end{bmatrix}\begin{Bmatrix} u_2 \\ u_3 \end{Bmatrix} \tag{1.2}$$

We want to form $\{F\} = [K]\{u\}$ for the entire structure by combining these individual element stiffnesses. Since there are three nodes and therefore three degrees of freedom for this structure, the force matrix $\{F\}$ will be of order 3×1, the stiffness matrix $[K]$ of order 3×3, and the displacement matrix $\{u\}$ of order 3×1.

The overall force-displacement relationship for the structure is constructed by placing the coefficients of u_1, u_2, and u_3 and corresponding F_{ij}'s into appropriate locations in the structural stiffness matrix $[K]$.

For example, the first column of $[k]$ in equation (1.1) multiplies u_1, thus

$$\begin{Bmatrix} F_{11} \\ - \\ - \end{Bmatrix} = \begin{bmatrix} k_1 & - & - \\ -k_1 & - & - \\ - & - & - \end{bmatrix} \begin{Bmatrix} u_1 \\ - \\ - \end{Bmatrix}$$

Next, the second column of equation (1.1) and the first column of equation (1.2) multiplies u_2 giving

$$\begin{Bmatrix} F_{11} \\ F_{21} + F_{22} \\ F_{32} \end{Bmatrix} = \begin{bmatrix} k_1 & -k_1 & - \\ -k_1 & k_1 + k_2 & - \\ - & -k_2 & - \end{bmatrix} \begin{Bmatrix} u_1 \\ u_2 \\ - \end{Bmatrix}$$

Finally, the second column of equation (1.2) multiplies u_3 giving

$$\begin{Bmatrix} F_{11} \\ F_{21} + F_{22} \\ F_{32} \end{Bmatrix} = \begin{bmatrix} k_1 & -k_1 & 0 \\ -k_1 & k_1 + k_2 & -k_2 \\ 0 & -k_2 & k_2 \end{bmatrix} \begin{Bmatrix} u_1 \\ u_2 \\ u_3 \end{Bmatrix}$$

Now, $F_1 = F_{11}$, $F_2 = F_{21} + F_{22}$, and $F_3 = F_{32}$ from the previous equilibrium equations. The final structural stiffness formulation becomes

$$\begin{Bmatrix} F_1 \\ F_2 \\ F_3 \end{Bmatrix} = \begin{bmatrix} k_1 & -k_1 & 0 \\ -k_1 & k_1 + k_2 & -k_2 \\ 0 & -k_2 & k_2 \end{bmatrix} \begin{Bmatrix} u_1 \\ u_2 \\ u_3 \end{Bmatrix}$$

Also, remember that F_1, F_2, and F_3 are either reactions or nodal applied loads. Naturally, this structural stiffness matrix is identical to the one obtained earlier for the same two-element structure by direct means.

One procedure that can automate the construction of the structural stiffness matrix from the individual elemental stiffness matrices involves identifying the rows and columns of these individual elemental stiffnesses with the global displacements associated with them. For example, using the stiffness matrices in equations (1.1) and (1.2) we have:

Member 1:

$$\begin{array}{c} \begin{matrix} \;\;1 & \;\;\;\;\;2 \end{matrix} \\ \begin{bmatrix} k_1 & -k_1 \\ -k_1 & k_1 \end{bmatrix} \begin{matrix} 1 \\ 2 \end{matrix} \end{array}$$

Member 2:

$$\begin{array}{c} \begin{matrix} \;\;2 & \;\;\;\;\;3 \end{matrix} \\ \begin{bmatrix} k_2 & -k_2 \\ -k_2 & k_2 \end{bmatrix} \begin{matrix} 2 \\ 3 \end{matrix} \end{array}$$

where the numbers identifying the rows and columns are those of the u displacements associated with each element. The structural stiffness matrix is then constructed in the following way:

(1) Create a square matrix that is of order equal to the total number of degrees of freedom of the structure (the structural stiffness matrix). In this case a 3×3 matrix is created.

(2) Place the elements of each individual stiffness matrix in the rows and columns of the new matrix corresponding to the global displacement coordinates.

(3) If there is more than one element to be placed in the same location in the overall structural stiffness matrix, the elements are added at that location.

For example, k_{22} in the structural stiffness matrix will be $k_1 + k_2$ since k_1 for member 1 and k_2 for member 2 both lie at global coordinates 2–2.

We can visually represent this assembly process by overlaying the individual elemental stiffness matrices on the structural stiffness matrix. Figure 1-10 illustrates this process.

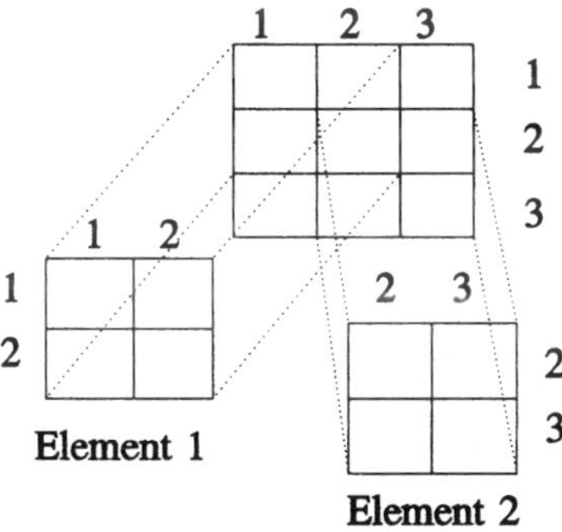

Figure 1-10 Combining elemental stiffnesses.

For the problem being considered, $u_1 = 0$, $F_2 = 25^k$, and $F_3 = -10^k$. The above equation becomes

$$\begin{Bmatrix} F_1 \\ 25^k \\ -10^k \end{Bmatrix} = \begin{bmatrix} k_1 & -k_1 & 0 \\ -k_1 & k_1 + k_2 & -k_2 \\ 0 & -k_2 & k_2 \end{bmatrix} \begin{Bmatrix} 0 \\ u_2 \\ u_3 \end{Bmatrix} \tag{1.3}$$

Note from this equation that we have two unknown displacements, u_2 and u_3. We next reduce this matrix equation to one that contains only the unknown displacements. We perform this task by noting that the first column of the matrix will multiply the known zero displacement and therefore can be eliminated. The first row can also be eliminated at this point since it will only be used to determine the reaction F_1 after the two unknown displacements have been found. Another way to think about this process is to notice that if we would expand this matrix equation, the last two equations would contain the applied nodal forces and the two unknown displacements. After eliminating the first row and column we have

$$\begin{Bmatrix} 25^k \\ -10^k \end{Bmatrix} = \begin{bmatrix} k_1 + k_2 & -k_2 \\ -k_2 & k_2 \end{bmatrix} \begin{Bmatrix} u_2 \\ u_3 \end{Bmatrix} \tag{1.4}$$

Since the elements of the stiffness matrix in equation (1.4) are known, this matrix equation represents two scalar equations in the two unknown displacements u_2 and u_3. The left hand side of equation (1.4) contains the applied nodal loads.

Let $E_1 = E_2 = 29 \times 10^3$ ksi, $A_1 = 2$ in^2, $A_2 = 1$ in^2, $L_1 = L_2 = 10$ in. Then $k_1 = 29 \times 10^3(2 \text{ in}^2)/10 \text{ in} = 58 \times 10^2$ k/in, and $k_2 = 29 \times 10^2$ k/in.

Equation 1.4 becomes

$$\begin{Bmatrix} 25^k \\ -10^k \end{Bmatrix} = \begin{bmatrix} 87 \times 10^2 & -29 \times 10^2 \\ -29 \times 10^2 & 29 \times 10^2 \end{bmatrix} \begin{Bmatrix} u_2 \\ u_3 \end{Bmatrix} \tag{1.5}$$

Solving equation (1.5) we find $u_2 = .002586$ in and $u_3 = -.000862$ in.

We determine the reaction F_1 by expanding the first row of equation (1.3). This yields $F_1 = -k_1u_2 = -15k$. The member forces are found by using u_2 and u_3 in equations (1.1) and (1.2).

$$\begin{Bmatrix} F_{11} \\ F_{21} \end{Bmatrix} = 58 \times 10^2 \begin{bmatrix} 1 & -1 \\ -1 & 1 \end{bmatrix} \begin{Bmatrix} 0 \\ .002586 \end{Bmatrix} = \begin{Bmatrix} -15^k \\ 15^k \end{Bmatrix}$$

$$\begin{Bmatrix} F_{22} \\ F_{32} \end{Bmatrix} = 29 \times 10^2 \begin{bmatrix} 1 & -1 \\ -1 & 1 \end{bmatrix} \begin{Bmatrix} .002586 \\ -.000862 \end{Bmatrix} = \begin{Bmatrix} 10^k \\ -10^k \end{Bmatrix}$$

Referring to Figure 1-11, the member free-body diagrams are shown with the actual directions of the member forces. Remember that positive member forces act to the right along the axis of the member.

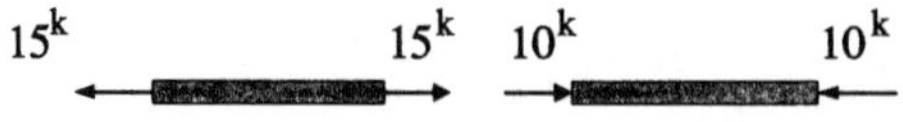

Figure 1-11 Member free-body diagrams.

To illustrate the process further, consider the following problem:

Example 1.2

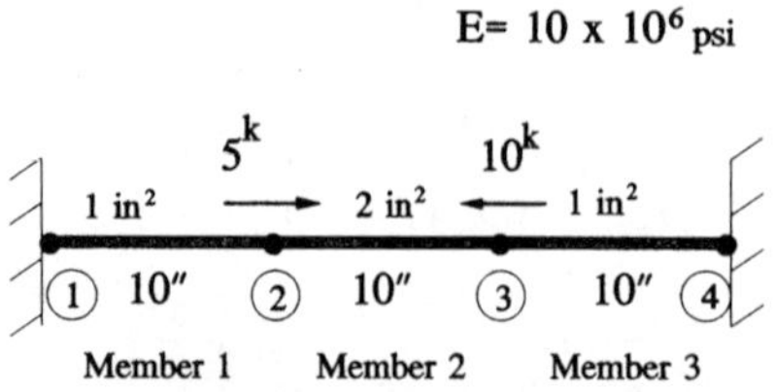

Figure 1-12

Using the material and geometric properties given, we first write the elemental force-displacement relationships.

Member 1:

$$\begin{Bmatrix} F_{11} \\ F_{21} \end{Bmatrix} = 10 \times 10^5 \begin{bmatrix} 1 & -1 \\ -1 & 1 \end{bmatrix} \begin{Bmatrix} u_1 \\ u_2 \end{Bmatrix} \begin{matrix} 1 \\ 2 \end{matrix} \quad \text{(columns 1, 2)}$$

Member 2:

$$\begin{Bmatrix} F_{22} \\ F_{32} \end{Bmatrix} = 10 \times 10^5 \begin{bmatrix} 2 & -2 \\ -2 & 2 \end{bmatrix} \begin{Bmatrix} u_2 \\ u_3 \end{Bmatrix} \begin{matrix} 2 \\ 3 \end{matrix} \quad \text{(columns 2, 3)}$$

Member 3:

$$\begin{Bmatrix} F_{33} \\ F_{43} \end{Bmatrix} = 10 \times 10^5 \begin{bmatrix} 1 & -1 \\ -1 & 1 \end{bmatrix} \begin{Bmatrix} u_3 \\ u_4 \end{Bmatrix} \begin{matrix} 3 \\ 4 \end{matrix} \quad \text{(columns 3, 4)}$$

Combining, using the labeled rows and columns, we have

$$\begin{Bmatrix} R_1 \\ 5^k \\ -10^k \\ R_4 \end{Bmatrix} = \begin{Bmatrix} F_{11} \\ F_{21}+F_{22} \\ F_{32}+F_{33} \\ F_{43} \end{Bmatrix} = 10\times 10^5 \begin{array}{c} \begin{array}{cccc} 1 & 2 & 3 & 4 \end{array} \\ \begin{bmatrix} 1 & -1 & 0 & 0 \\ -1 & 3 & -2 & 0 \\ 0 & -2 & 3 & -1 \\ 0 & 0 & -1 & 1 \end{bmatrix} \end{array} \begin{Bmatrix} u_1 \\ u_2 \\ u_3 \\ u_4 \end{Bmatrix} \begin{array}{c} 1 \\ 2 \\ 3 \\ 4 \end{array} \tag{1.6}$$

Since both u_1 and u_4 are zero because the ends of the structure are fixed, we eliminate rows 1 and 4 and columns 1 and 4 giving the following equation to be solved for the unknown displacements u_2 and u_3;

$$\begin{Bmatrix} 5^k \\ -10^k \end{Bmatrix} = \begin{bmatrix} 3 & -2 \\ -2 & 3 \end{bmatrix} \begin{Bmatrix} u_2 \\ u_3 \end{Bmatrix} \tag{1.7}$$

This equation is called the *reduced structural stiffness* equation. Solving, we find $u_2 = -.001$ in, and $u_3 = -.004$ in. From the complete set of equations we next find the reactions R_1 and R_4. $R_1 = 10\times 10^5(u_1 - u_2) = 1000$ lb $= 1^k$. $R_4 = 10\times 10^5(-u_3 + u_4) = 4000$ lb $= 4^k$. Note that these reactions create overall equilibrium with the applied loads (see Figure 1-13).

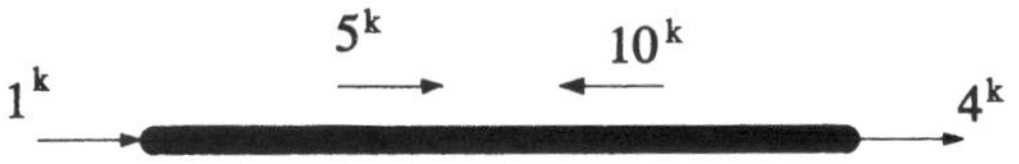

5^k 10^k 1^k 4^k

Figure 1-13 Applied loads and reactions.

We next determine the member forces.

Member 1:

$$\begin{Bmatrix} F_{11} \\ F_{21} \end{Bmatrix} = 10\times 10^5 \begin{bmatrix} 1 & -1 \\ -1 & 1 \end{bmatrix} \begin{Bmatrix} 0 \\ -.001 \end{Bmatrix} = \begin{Bmatrix} 1000\# \\ -1000\# \end{Bmatrix}$$

Member 2:

$$\begin{Bmatrix} F_{22} \\ F_{32} \end{Bmatrix} = 10\times 10^5 \begin{bmatrix} 2 & -2 \\ -2 & 2 \end{bmatrix} \begin{Bmatrix} -.001 \\ -.004 \end{Bmatrix} = \begin{Bmatrix} 6000\# \\ -6000\# \end{Bmatrix}$$

Member 3:

$$\begin{Bmatrix} F_{33} \\ F_{43} \end{Bmatrix} = 10\times 10^5 \begin{bmatrix} 1 & -1 \\ -1 & 1 \end{bmatrix} \begin{Bmatrix} -.004 \\ 0 \end{Bmatrix} = \begin{Bmatrix} 4000\# \\ -4000\# \end{Bmatrix}$$

Figure 1-14 shows free-body diagrams of the nodes and members.

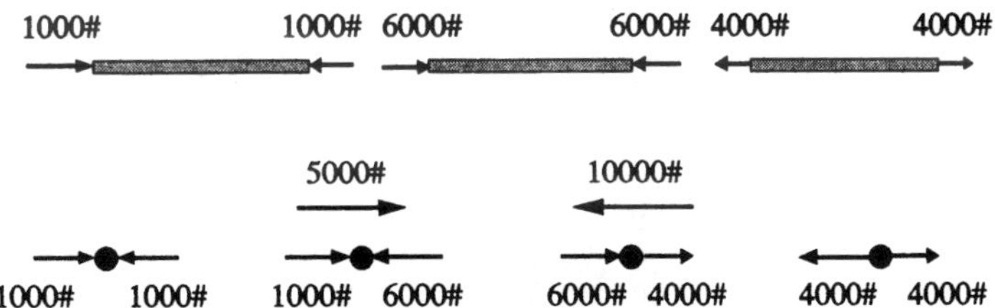

Figure 1-14 Free-body diagrams of members and nodes.

We could also obtain our reduced structural stiffness equation, (1.7) from the combined stiffness equation, (1.6), by using a partitioning technique on equation (1.6).

Consider rewriting equation (1.6) in the following way:

$$\begin{Bmatrix} F_p \\ F_s \end{Bmatrix} = \begin{bmatrix} K_{pp} & K_{ps} \\ K_{sp} & K_{ss} \end{bmatrix} \begin{Bmatrix} u_p \\ u_s \end{Bmatrix} \tag{1.8}$$

where

F_p = submatrix of applied loads
F_s = submatrix of reactions
u_p = submatrix of unknown displacements
u_s = submatrix of known displacements, including support movements.

Expanding equation (1.8) we have

$$F_p = K_{pp}u_p + K_{ps}u_s \tag{1.9}$$

$$F_s = K_{sp}u_p + K_{ss}u_s \tag{1.10}$$

Note that the number of applied loads and unknown displacements are always equal. Thus, K_{pp} will always be a square matrix.

From equation 1.9

$$u_p = K_{pp}^{-1}(F_p - K_{ps}u_s) \tag{1.11}$$

We can solve equation (1.11) for the unknown displacements. If reactions are desired, equation (1.10) can then be used.

For the previous example we rearrange equation (1.6) in the following way:

$$\begin{Bmatrix} 5^k \\ -10^k \\ R_1 \\ R_4 \end{Bmatrix} = 10 \times 10^5 \begin{bmatrix} 3 & -2 & -1 & 0 \\ -2 & 3 & 0 & -1 \\ -1 & 0 & 1 & 0 \\ 0 & -1 & 0 & 1 \end{bmatrix} \begin{Bmatrix} u_2 \\ u_3 \\ u_1 \\ u_4 \end{Bmatrix} \tag{1.12}$$

Referring to equation 1.8 we have

$$\{F_p\} = \begin{Bmatrix} 5^k \\ -10^k \end{Bmatrix}, \quad \{F_s\} = \begin{Bmatrix} R_1 \\ R_4 \end{Bmatrix}$$

$$[K_{pp}] = \begin{bmatrix} 3 & -2 \\ -2 & 3 \end{bmatrix}, \quad [K_{ps}] = \begin{bmatrix} -1 & 0 \\ 0 & -1 \end{bmatrix}$$

$$[K_{sp}] = \begin{bmatrix} -1 & 0 \\ 0 & -1 \end{bmatrix}, \quad [K_{ss}] = \begin{bmatrix} 1 & 0 \\ 0 & 1 \end{bmatrix}$$

$$\{u_p\} = \begin{Bmatrix} u_2 \\ u_3 \end{Bmatrix}, \quad \{u_s\} = \begin{Bmatrix} u_1 \\ u_4 \end{Bmatrix} = \begin{Bmatrix} 0 \\ 0 \end{Bmatrix}$$

From equation 1.11, since $\{u_s\}$ is null, we have $\{u_p\} = [K_{pp}]^{-1}\{F_p\}$ or

$$\begin{Bmatrix} u_2 \\ u_3 \end{Bmatrix} = \begin{bmatrix} 3 & -2 \\ -2 & 3 \end{bmatrix}^{-1} \begin{Bmatrix} 5^k \\ -10^k \end{Bmatrix} \tag{1.13}$$

Comparison of equation (1.13) with equation (1.7) shows that the solution for u_2 and u_3 will be identical.

1.6 STRUCTURES WITH SPECIFIED NON-ZERO DISPLACEMENTS

Consider Example 1.1 presented in section 1.5.

Instead of applying the 10k load at node 3, let us specify a displacement $u_3 = -.000862$ in (the value calculated previously).

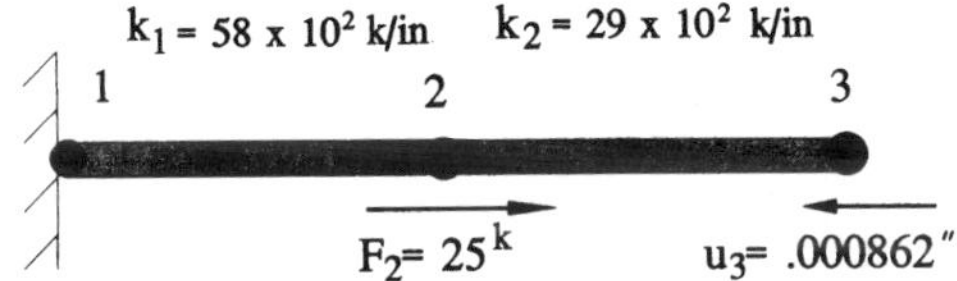

Element 1 Element 2 **Figure 1-15**

Our matrix equation for the structure now becomes

$$\begin{Bmatrix} R_1 \\ 25^k \\ R_3 \end{Bmatrix} = \begin{bmatrix} k_1 & -k_1 & 0 \\ -k_1 & k_1 + k_2 & -k_2 \\ 0 & -k_2 & k_2 \end{bmatrix} \begin{Bmatrix} 0 \\ u_2 \\ -.000862 \end{Bmatrix} \tag{1.14}$$

For this simple problem we can expand the second row of this equation to obtain

$$25^k = (k_1 + k_2)u_2 - k_2(-.000862).$$

Solving for u_2 yields $u_2 = .002586$ in as obtained previously.

If we use the partitioning technique presented in section 1.5, we reorder equation 1.14 as shown below.

$$\begin{Bmatrix} 25^k \\ R_2 \\ R_3 \end{Bmatrix} = \begin{bmatrix} k_1 + k_2 & -k_1 & -k_2 \\ -k_1 & k_1 & 0 \\ -k_2 & 0 & k_2 \end{bmatrix} \begin{Bmatrix} u_2 \\ 0 \\ -.000862 \end{Bmatrix}$$

Thus,

$$K_{pp} = [k_1 + k_2]\ , \qquad K_{ps} = [-k_1 \quad -k_2]$$

$$K_{sp} = \begin{Bmatrix} -k_1 \\ -k_2 \end{Bmatrix}\ , \qquad K_{ss} = \begin{bmatrix} k_1 & 0 \\ 0 & k_2 \end{bmatrix}$$

$$u_p = [u_2]\ , \qquad u_s = \begin{Bmatrix} 0 \\ -.000862 \end{Bmatrix}\ , \qquad F_p = [25^k]$$

From equation 1.11 we have

$$[u_2] = \frac{1}{k_1 + k_2}\left([25^k] - [-k_1 \quad -k_2]\begin{Bmatrix} 0 \\ -.000862 \end{Bmatrix}\right)$$

$$= \frac{1}{k_1 + k_2}(25^k - k_2[0.000862])$$

$$= \frac{1}{87 \times 10^2}(25^k - 2.5^k) = .0025861 \text{ in}$$

Of course, this is the solution obtained previously.

There is also a numerical procedure that can be automated and is capable of treating complex problems. Since many matrix inversion routines make use of the symmetry of the global stiffness matrix, it would be beneficial for our procedure to maintain this symmetry. The following method for handling non-zero displacements does this.

(1) Eliminate the rows and columns corresponding to zero displacements as before. For the previous example we obtain

$$\begin{Bmatrix} 25^k \\ R_3 \end{Bmatrix} = \begin{bmatrix} k_1 + k_2 & -k_2 \\ -k_2 & k_2 \end{bmatrix}\begin{Bmatrix} u_2 \\ -.000862 \end{Bmatrix}$$

(2) If a displacement u_k is specified at coordinate n, multiply k_{nn} by a large number M and replace the force value in row n by $u_k \times M \times k_{nn}$. We have

$$\begin{Bmatrix} 25^k \\ -.000862 \times k_2 \times M \end{Bmatrix} = \begin{bmatrix} k_1 + k_2 & -k_2 \\ -k_2 & M \times k_2 \end{bmatrix}\begin{Bmatrix} u_2 \\ u_3 \end{Bmatrix} \qquad (1.15)$$

(3) Solve for the displacements in the normal fashion.

To illustrate with this example, we invert the modified stiffness matrix in equation (1.15) to obtain

$$[K]^{-1} = \begin{bmatrix} Mk_2/(Mk_2(k_1 + k_2) - k_2^2) & k_2/(\text{same denom.}) \\ k_2/(\text{same denom.}) & (k_1 + k_2)/(\text{same denom.}) \end{bmatrix}$$

Premultiply the force-displacement equation by $[K]^{-1}$ in order to solve for u_2 and u_3. We find:

$$u_2 = \frac{25k_2 - .000862k_2^2}{k_1k_2 + k_2^2 - k_2^2/M}$$

$$u_3 = \frac{(25k_2/M) - .000862(k_1k_2 + k_2^2)}{k_1k_2 + k_2^2 - k_2^2/M}$$

In these equations, the last term in the denominator is very small due to the presence of the large number M. This term is neglected in comparison to the other terms. In addition, the first term in the numerator of the expression for u_3 is also neglected in comparison to the other terms. Substituting the values for k_1 and k_2 of 58×10^2 ksi and 29 × 10² ksi respectively, we find that $u_2 = .002586$ in and $u_3 = -.000862$ in (the specified displacement).

1.7 NON-NODAL FORCES

Up to this point we have dealt with structures that have forces applied only at the nodes. Clearly, we need to be able to address problems that have concentrated and distributed loads applied between nodal points. Figure 1-16 shows such a structure.

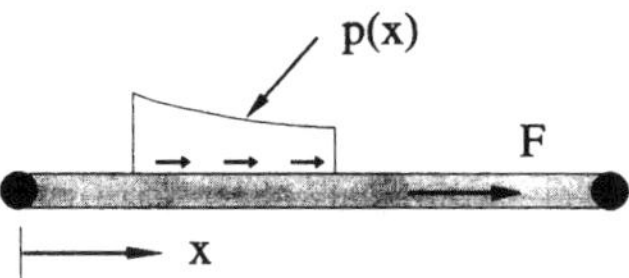

Figure 1-16 Non-nodal loads on an element.

Of course, one approach that could be used to solve this problem would be to add several more nodal points—one at the concentrated load so that the load will be acting at a nodal point location, and several more in the region of the distributed load. We would then break up the distributed load into a series of concentrated forces acting on the additional nodal points in that region. The magnitude of the concentrated forces would generally be based on tributary length. For example, in Figure 1-17 the portion of the distributed load assigned to each node is indicated. The total load applied to a node would be that load contained in a length half the distance to each adjacent node.

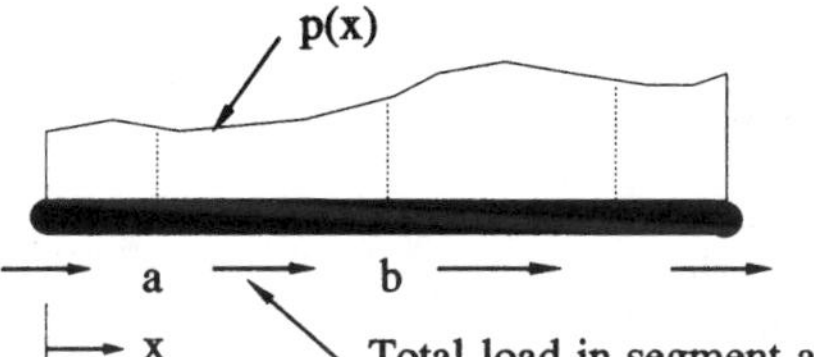

Figure 1-17 Tributary lengths.

The number of additional nodes required would be dependent on how rapidly the distributed load changes, that is, the gradient of the load. The faster the magnitude of the load changes, the more nodal points necessary to achieve a reasonable approximate solution to the problem.

A major difficulty arises, however, when using this technique. Since we have added nodes we have added degrees of freedom and consequently increased the size of our force, stiffness, and displacement matrices. As a result, solution time and storage requirements will increase. We need to find a method that will enable us to determine nodal forces at the ends of the bar that are equivalent to the loads applied between nodes. The use of fixed end forces allows us to perform this task.

Suppose we fix each node in position before applying the loads. After the loads have been applied, forces are required at the nodes in order to maintain zero displacements. These are the *fixed end forces*. If the nodal restraints are now removed, the ensuing displacements will be those caused by loads equivalent to the opposite of the fixed end forces. That is, the fixed end forces are removed when the nodes are allowed to displace. Thus the displacements will be due to loads equal in magnitude to the fixed end forces but opposite in sense. Therefore, equivalent forces that must be applied to the nodes of an element in order to account for non-nodal forces are exactly opposite of the fixed end forces.

Consider the superposition diagram shown in Figure 1.18.

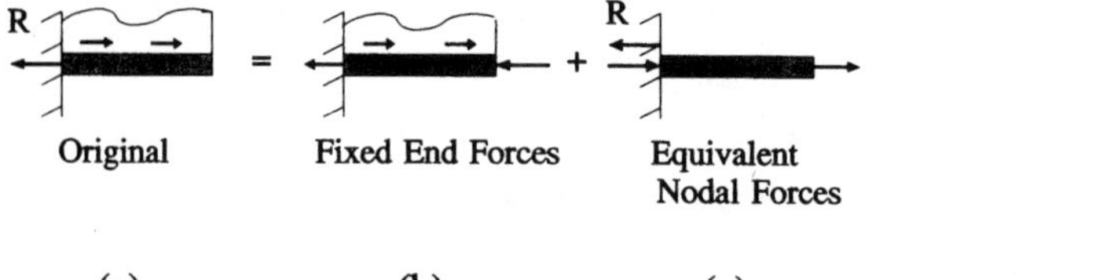

Figure 1-18 Superposition diagram.

For determining the nodal displacements we solve part (c) of the superposition diagram shown in Figure 1-18. Since part (b) of the diagram has zero nodal displacements, the displacements found in part (c), using the equivalent nodal forces, are the true nodal displacements of the original structure. However, we must add the forces in both parts (b) and (c) in order to obtain the forces acting at the nodes of the original structure.

Formulating the procedure algebraically, we can write

$$\{F\} + \{F_{equiv}\} = [K]\{u\} \tag{1.16}$$

The solution of equation (1.16) yields the true nodal displacements.

The true nodal forces acting on the members are found by using

$$F_i = [k]_i\,\{u\}_i - \{F_{equiv}\} = [k]_i\,\{u\}_i + \{F_{fixed\text{-}end}\}_i$$

where all subscripted terms refer to the element being considered.

After determining the true nodal forces we can then construct the axial force diagrams by considering free-body diagrams of the original structure with the actual loads applied.

The determination of fixed end forces and the use of equivalent nodal forces are illustrated in the following examples.

Example 1.3

Determine the equivalent nodal forces for the uniformly distributed load shown. The cross-sectional area of the bar is constant. We shall determine the fixed end forces and then reverse their sense.

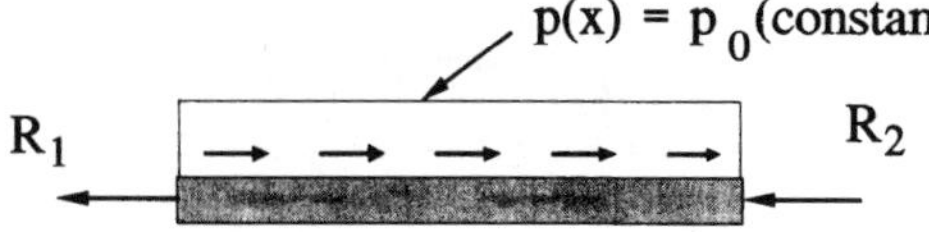

Figure E1-3a Bar element with uniformly distributed load.

Since we have two unknown axial forces, R_1 and R_2, and only one axial equilibrium equation, the structure has one degree of indeterminacy. The compatibility condition we need to generate in order to obtain an additional equation results from the knowledge that there is no relative displacement of one end of the member with respect to the other since both nodes are fixed.

Drawing a free-body diagram of a portion of the structure we have

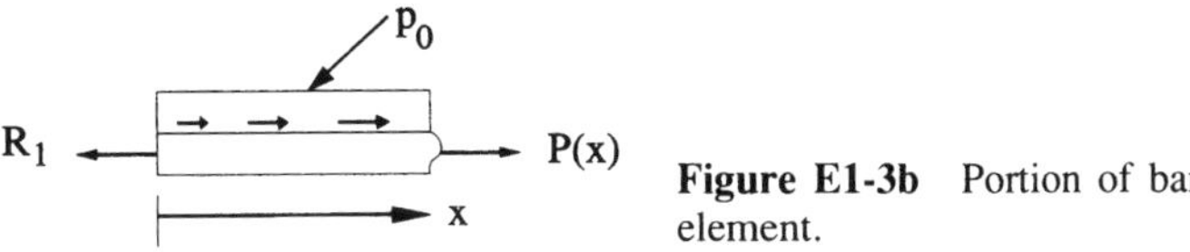

Figure E1-3b Portion of bar element.

from which $P(x) = R_1 - p_0 x$. The strain at position x is $P(x)/AE$. The total elongation of the bar is zero, and is obtained by integrating the strain over the length of the bar. Thus,

$$0 = (1/AE)\int_0^l (R_1 - p_0 x)dx = (1/AE)\left[R_1 l - p_0 l^2/2\right]$$

from which $R_1 = p_0 l/2$.

From overall equilibrium, $R_1 + R_2 = p_0 l$. Thus, $R_2 = p_0 l/2$.

Since both R_1 and R_2 are positive, we have assumed the correct directions for these forces. Reversing their directions yields the equivalent nodal loads.

Note that each equivalent force is equal to one-half the total load.

$p_0\, l/2$ $p_0\, l/2$

Figure E1-3c Equivalent nodal loads.

Example 1.4

Solve for the nodal displacements and member end forces for the structure shown in Figure E1-4a. Construct the axial force diagrams for each member.

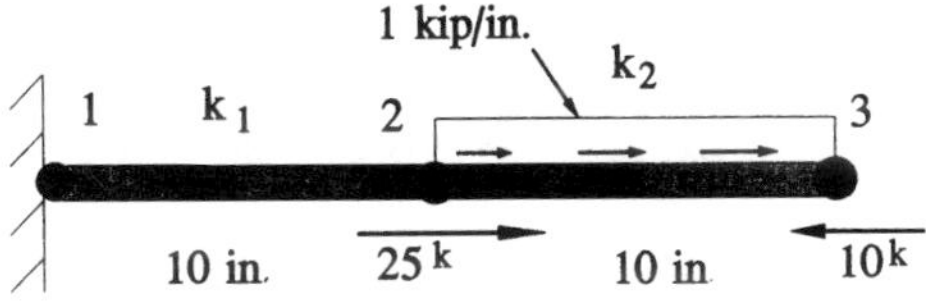

Element 1 Element 2

Figure E1-4a Example 1.4.

Since the uniformly distributed load totals 10 kips, 5 kips acting in the positive direction (to the right) are added to the applied nodal forces. This yields the nodal forces shown in Figure E1-4b.

Using values of $k_1 = 58 \times 10^2$ k/in and $k_2 = 29 \times 10^2$ k/in, and equation (1.16), the structural equation becomes

$$\begin{Bmatrix} R_1 \\ 30^k \\ -5^k \end{Bmatrix} = \begin{bmatrix} 58 \times 10^2 & -58 \times 10^2 & 0 \\ -58 \times 10^2 & 87 \times 10^2 & -29 \times 10^2 \\ 0 & -29 \times 10^2 & 29 \times 10^2 \end{bmatrix} \begin{Bmatrix} 0 \\ u_2 \\ u_3 \end{Bmatrix}$$

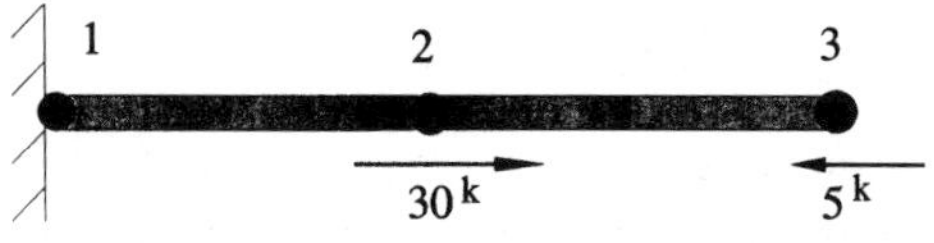

Element 1 Element 2

Figure E1-4b Total nodal loads.

Reducing the above equation by eliminating the first row and column, which correspond to the reaction and zero displacement at the fixed end, we solve for the unknown displacements obtaining $u_2 = .00431$ in and $u_3 = .00258$ in.

Solving for member forces we have

$$\begin{Bmatrix} F_{11} \\ F_{21} \end{Bmatrix} = 58 \times 10^2 \begin{bmatrix} 1 & -1 \\ -1 & 1 \end{bmatrix} \begin{Bmatrix} 0 \\ .00431 \end{Bmatrix} + \begin{Bmatrix} 0 \\ 0 \end{Bmatrix}$$

Thus, $F_{11} = -25$ kips, $F_{21} = +25$ kips.

$$\begin{Bmatrix} F_{22} \\ F_{32} \end{Bmatrix} = 29 \times 10^2 \begin{bmatrix} 1 & -1 \\ -1 & 1 \end{bmatrix} \begin{Bmatrix} .00431 \\ .00258 \end{Bmatrix} + \begin{Bmatrix} -5 \\ -5 \end{Bmatrix}$$

Thus, $F_{22} = 0$, $F_{32} = -10$ kips.

Note that the last terms in the above equations reflect the addition of the fixed end forces (opposite of the equivalent nodal loads).

Figure E1-4c shows free-body diagrams of the nodes and members, and Figure E1-4d shows the axial force diagrams.

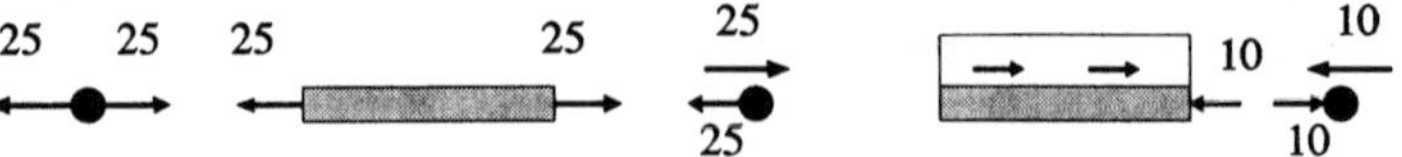

Figure E1-4c Free-body diagrams of members and nodes.

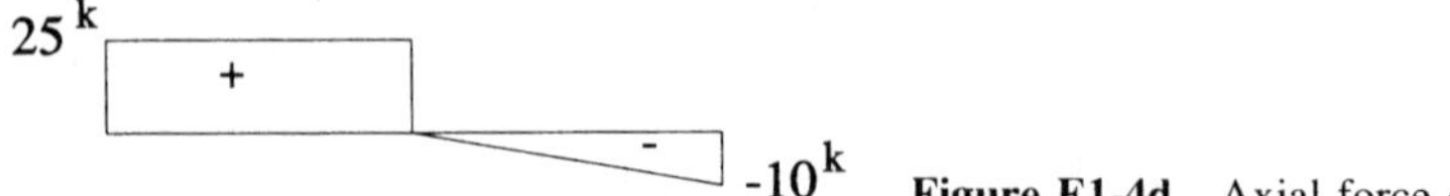

Figure E1-4d Axial force diagram.

Before constructing the axial force diagrams, free body diagrams such as those of Figure E1-4c should be drawn with the forces acting in the true directions. Remember that our matrix solution sign convention requires positive forces to act in the positive direction of the axis of the member independent of the end being considered. This is different from the standard strength of materials sign convention where tension in the member is considered positive. By constructing the free-body diagrams we can easily relate the actual force directions to the strength of materials sign convention for forces.

Another very important reason for drawing the free body diagrams is that it gives us the ability to check equilibrium of the members and nodes. This is essential with any output from a computer program, since errors in input or the program itself can be readily found.

1.8 THERMAL EFFECTS

When a temperature change occurs in a bar element, an axial strain linearly proportional to this temperature change takes place. This strain is generally expressed as $\epsilon_T = \alpha(\Delta T)$, where α is the coefficient of thermal expansion and has units of strain/°F or strain/°C, and ΔT is the temperature change. If the bar has a length L, the total change in length due to a temperature change over the entire bar is $\Delta L = \epsilon_T L = \alpha(\Delta T)L$. We can treat thermal effects in rods in the same way that we dealt with non-nodal forces. That is, we will use the concept of fixed end forces and equivalent nodal loads.

As in problems dealing with non-nodal loads, the equivalent nodal forces are the same magnitude as the fixed end forces but opposite in sense. After applying the equivalent nodal loads and solving for the unknown nodal displacements, we find the member forces by superposing the fixed end forces and the forces calculated using the nodal displacements. Since $\sigma = P/A = E\epsilon = E\alpha(\Delta T)$, then $P = EA\alpha(\Delta T)$ and the fixed end forces for the rod are given by

$$\{F\}_{fixed} = EA\alpha(\Delta T)\begin{Bmatrix} 1 \\ -1 \end{Bmatrix}$$

Of course, since we are dealing exclusively with linear systems, the superposition principle is valid and the thermal effects can occur simultaneously with both applied nodal loads and non-nodal loads.

Example 1.5

For the structure shown in Figure E1-5a, determine the nodal displacements and member forces. In addition to the load applied at node 2, member 2 undergoes a temperature increase of 60°F. All bar areas are 2 in^2, $E = 29 \times 10^6$ psi, and $\alpha = 6.5 \times 10^{-6}$ in/in/°F.

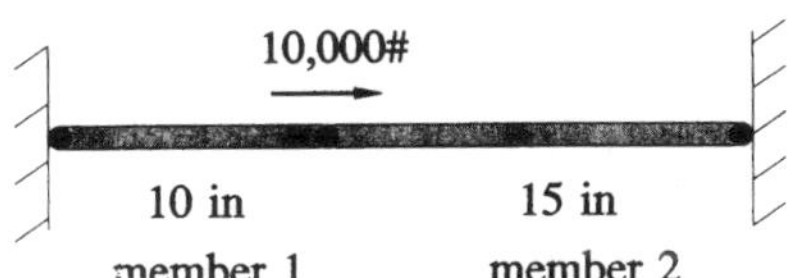

Figure E1-5a Example 1.5.

The fixed-end forces due to the temperature change in member 2 are found using $P = EA\Delta/L = (EA/L)\alpha(\Delta T)L = EA\alpha(\Delta T) = 22{,}620$ lbs compression. The equivalent nodal forces (opposite of the fixed end forces) are shown in Figure E1-5b.

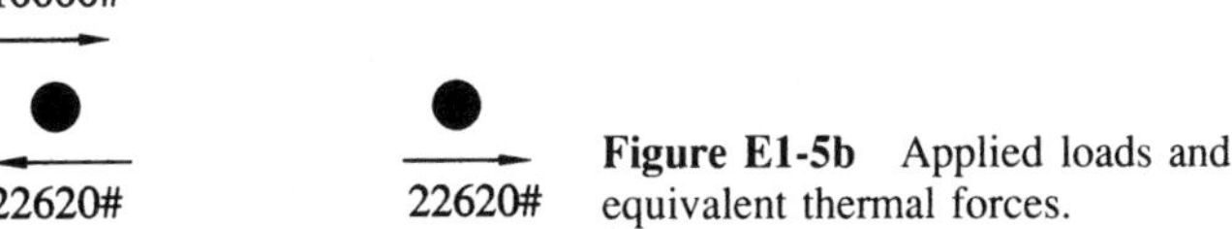

Figure E1-5b Applied loads and equivalent thermal forces.

For member 1:

$$\begin{Bmatrix} F_1 \\ F_2 \end{Bmatrix} = 5.8 \times 10^6 \begin{bmatrix} 1 & -1 \\ -1 & 1 \end{bmatrix} \begin{Bmatrix} u_1 \\ u_2 \end{Bmatrix}$$

For member 2:

$$\begin{Bmatrix} F_2 \\ F_3 \end{Bmatrix} = 3.866 \times 10^6 \begin{bmatrix} 1 & -1 \\ -1 & 1 \end{bmatrix} \begin{Bmatrix} u_2 \\ u_3 \end{Bmatrix}$$

Combining,

$$\begin{Bmatrix} F_1 \\ F_2 \\ F_3 \end{Bmatrix} = \begin{Bmatrix} F_1 \\ -12620 \\ F_3 \end{Bmatrix} = \begin{bmatrix} 5.8 \times 10^6 & -5.8 \times 10^6 & 0 \\ -5.8 \times 10^6 & 9.666 \times 10^6 & -3.866 \times 10^6 \\ 0 & -3.866 \times 10^6 & 3.866 \times 10^6 \end{bmatrix} \begin{Bmatrix} u_1 \\ u_2 \\ u_3 \end{Bmatrix}$$

Since u_1 and u_3 are both zero, the second equation yields $-12{,}620\# = 9.6666 \times 10^6 u_2$. Thus, $u_2 = -.0013055$ in.

Member forces:

Member 1:

$$\begin{Bmatrix} F_1 \\ F_2 \end{Bmatrix} = 5.8 \times 10^6 \begin{bmatrix} 1 & -1 \\ -1 & 1 \end{bmatrix} \begin{Bmatrix} 0 \\ -.0013055 \end{Bmatrix} = \begin{Bmatrix} 7572 \\ -7572 \end{Bmatrix} \text{ pounds}$$

Member 2:

$$\begin{Bmatrix} F_2 \\ F_3 \end{Bmatrix} = 3.866 \times 10^6 \begin{bmatrix} 1 & -1 \\ -1 & 1 \end{bmatrix} \begin{Bmatrix} -.0013055 \\ 0 \end{Bmatrix} + \begin{Bmatrix} 22620 \\ -22620 \end{Bmatrix} = \begin{Bmatrix} 17572 \\ -17572 \end{Bmatrix} \text{ pounds}$$

Figure E1-5c shows free-body diagrams of the members and node 2.

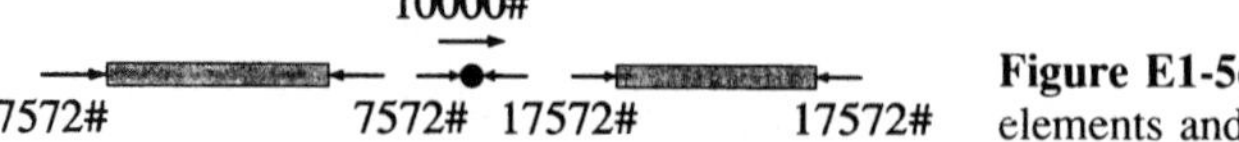

Figure E1-5c Free-body diagrams of elements and nodes.

Note that the member nodes are in equilibrium. This should always be checked.

1.9 COMPUTER FORMULATION

To automate the solution process illustrated above, we need to be able to combine the individual stiffness matrices into the structural stiffness matrix. In addition, we need to reduce the resulting matrix using the known displacements to solve for the unknown displacements. A general outline for a computer program to solve these one-dimensional problems is cited below.

(1) Sketch the structure labeling nodes and members and show applied loads.

(2) Enter geometric and material properties, applied loads, and support conditions (known displacements).

(3) For the number of elements, generate each elemental stiffness matrix and place its elements in the appropriate locations in the global (structural) stiffness matrix.

(4) Reduce the global stiffness matrix using the boundary conditions (including known non-zero displacements).

(5) Solve the resulting equations for the unknown displacements.

(6) Calculate the member forces using the computed displacements and the individual member stiffness matrices.

(7) Print out input data, member forces, and nodal displacements.

Let us assume that the following data have been entered into the program:

```
NE = number of elements
NN = number of nodes
```

For each member:

```
E(I)  = modulus of elasticity of member I
A(I)  = cross-sectional area of member I
L(I)  = length of member I
NL(I)  = left node number of member I
NR(I)  = right node number of member I
```

We want to form SK(I,J), the global structural stiffness matrix. Consider the following code fragment:

```
FOR I = 1 TO NN
FOR J = 1 TO NN
SK(I,J)=0.  [zero structural stiffness matrix]
NEXT J: NEXT I
FOR I=1 TO NE  [loop on number of elements]
EK=E(I)*A(I)/L(I)
EKT(1,1)=EKT(2,2)=EK
EKT(1,2)=EKT(2,1)= -EK  [elements of member stiffness matrix]
IJ(1)=NL(I)
IJ(2)=NR(I)  [counters(left and right node numbers)]
FOR IR = 1 TO 2  [row index]
FOR IC = 1 TO 2  [column index]
KR = IJ(IR)  [row of SK matrix]
KC = IJ(IC)  [column of SK matrix]
SK(KR,KC) = SK(KR,KC) + EKT(IR,IC)  [accumulate element stiffnesses in
global stiffness matrix]
NEXT IC:NEXT IR
NEXT I
```

For each elemental stiffness matrix:

(a) elements 1,1 and 2,1 multiply the left node number displacement;

(b) elements 1,2 and 2,2 multiply the right node number displacement.

Thus, element 1,1 should be placed in the SK matrix in both the row and column corresponding to the left node number global displacement. Also, element 1,2 should be placed in the row corresponding to the left node number and the column corresponding to the right node number. Element 2,2 should be placed in the row and column corresponding to the right node number. Element 2,1 should be placed in the row corresponding to the right node number and the column corresponding to the left node number.

The code fragment shown above accomplishes these tasks. Note that the elemental stiffness matrices are not saved. Each matrix is calculated and immediately placed in the global stiffness matrix. Although this will require recalculation of the elemental stiffnesses in order to find member forces, it reduces memory requirements.

Next, consider the reduction of the global stiffness and force matrices using the boundary conditions.

If the displacement boundary conditions are all homogeneous (i.e., $u_k = 0$), as in the previous examples, we must eliminate the rows and columns corresponding to the zero displacements. The rows eliminated correspond to the unknown reactive forces, and the columns correspond to the zero displacements (the same number row and column for each zero displacement).

Clearly, we must specify which nodes have zero displacement when entering data. Suppose that for each node we have specified a restraint code where zero (0) means that the node is free to displace and one (1) means that the node has zero displacement. We have stored these values in an array KRES(I), where I ranges from 1 to the number of nodes. In Example 1.1 the KRES(I) matrix would be {1 0 0} since we have three nodes and node number 1 has zero displacement.

We first determine the order of the reduced stiffness matrix SKR(I,J). The order of this matrix will be equal to the number of degrees of freedom minus the number of nodes that are restrained. Remember that for this one-dimensional element, each node has only one degree of freedom. A code fragment to find the order of SKR might look like the following

```
KSUM=0  [initialize KSUM to zero]
FOR I=1 TO NN  [loop on number of nodes]
IF(KRES(I))>0 THEN KSUM=KSUM+1  [accumulate the number of restraints]
NEXT I
NKR=NN-KSUM  [calculate the order of SKR]
```

We next determine the numbers of the rows and columns that will be kept.

```
J=0
FOR I=1 TO NN
IF KRES(I)=0 THEN J=J+1:KEPT(J)=I
NEXT I
```

In Example 1.1, KEPT(J)={2 3}, which are the numbers of the rows and columns we want to keep when constructing the reduced stiffness matrix SKR.

Now we fill SKR and generate a new reduced force matrix that contains the known applied loads. These will be the forces of the original force matrix that are in the rows corresponding to the row numbers kept.

```
FOR I=1 TO NKR
N=KEPT(I): FR(I)=F(N)
FOR J=1 TO NKR
M=KEPT(J): SKR(I,J)=SK(N,M)
NEXT J: NEXT I
```

The solution of the resulting equations requires the inversion of SKR and multiplication of the reduced force matrix FR. That is, since {FR}=[SKR]{UR}, then {UR}= $[\text{SKR}]^{-1}$\{FR\} where {UR} is the matrix of calculated unknown displacements. We must, however, place the calculated displacements in the correct row of the global displacement matrix $\{u\}$, which contains all displacements, including the ones that are zero or specified. This can be accomplished as illustrated below, where it is assumed that after inversion of SKR, SKR contains the inverse. In other words, the original elements of SKR have been replaced by its own inverse during the process of inversion. In the following code fragment, U represents the unknown displacements to be calculated and DU represents the complete global displacement matrix. After inverting SKR,

```
FOR I=1 TO NKR: U(I)=0: NEXT I  [zero all unknown displacements]
FOR I=1 TO NKR
FOR J=1 TO NKR
U(I)=U(I)+SKR(I,J)*FR(J)    [calculate the unknown displacements]
NEXT J: NEXT I
FOR I=1 TO NN: DU(I)=0.: NEXT I  [zero global displacements]
FOR I=1 TO NKR: ND=KEPT(I): DU(ND)=U(I): NEXT I
```

Note that the code fragment above assumes that there are no non-zero specified displacements. In addition, we have included forces applied at only the nodes.

Next, we calculate the member forces. Note that we can compute the values of the unknown reactions from the original SK matrix and DU displacements, or by using the member forces at the nodes with zero displacements.

In calculating the member forces, we need only the force at one end of the member since the force is constant throughout the bar when loads are applied only at the nodes. We choose the right end of the member since a positive force at this end indicates tension in the member, which is consistent with the common strength of materials sign convention.

For element j, with nodes i and $i+1$, $F_j = k_j(u_{i+1} - u_i)$. The code becomes

```
FOR I=1 TO NE  [loop on number of elements]
P(I)=(E(I)*A(I)/L(I))*(DU(I+1)-DU(I))
NEXT I
```

P(I), of course, represents the force in member I.

We can now print out all displacements and forces. In general, it is always good procedure to print out all data that was entered into the program as well. If this is done, then it becomes much easier to check input data and, if necessary, to debug the program.

1.10 SUMMARY

In this chapter we have derived the elemental stiffness matrix for a one-dimensional bar. We accomplished this by using the basic definition of stiffness as the force at i corresponding to a unit displacement at j. We showed how the individual elemental stiffnesses are combined in order to form the structural stiffness matrix. A method of treating specified non-zero displacements was presented. Equivalent nodal forces due to non-nodal loads were derived by considering the fixed end forces corresponding to these non-nodal loads. We saw that these equivalent nodal forces resulted in the correct nodal displacements; however, the member forces required the addition of the fixed end forces in order to obtain the true end forces. We also outlined the programming steps necessary to develop a computer program for solution of these one-dimensional problems, and we presented algorithms to accomplish this task.

In the chapters that follow we will be investigating the use of many commonly used elements for solving structural problems. Later in the text we shall present alternative ways of deriving the elemental stiffness matrices, the overall structural stiffness matrix, and the fixed end forces due to non-nodal loads.

PROBLEMS

1.1 For the structure shown in Figure P1-1

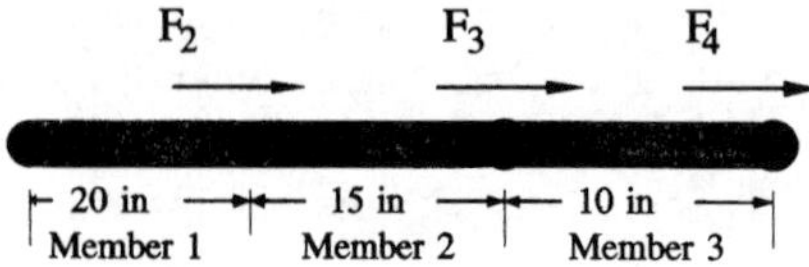

Figure P1-1

(1) Determine the structural stiffness matrix for the entire structure by using the basic definition of stiffness.
(2) Determine the structural stiffness matrix by combining individual elemental stiffnesses.
(3) If the left end of the structure is fixed, solve for the nodal displacements and member forces if $F_2 = 10$ kips, $F_3 = -5$ kips, and $F_4 = 8$ kips. Use the partitioning method and the method of removing rows and columns from the original structural stiffness matrix. $E = 29 \times 10^6$ psi for all members.

1.2 For the structure shown in Figure P1-2, solve for nodal displacements and member forces.

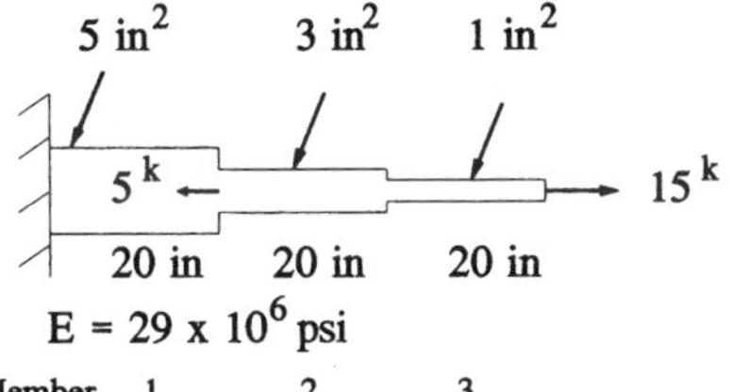

Figure P1-2

1.3 Solve for nodal displacements and member end forces for the structure in Figure P1-3. Also construct the axial force diagrams.

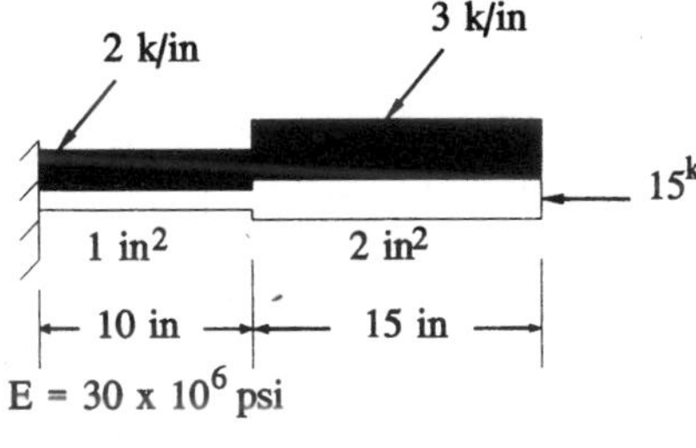

Figure P1-3

1.4 Referring to Example 1.4, replace the 10 kip load at node 3 with a specified displacement of +.00258 in. Solve for all displacements and member forces. Use the partioning method and the numerical method presented in section 1.6.

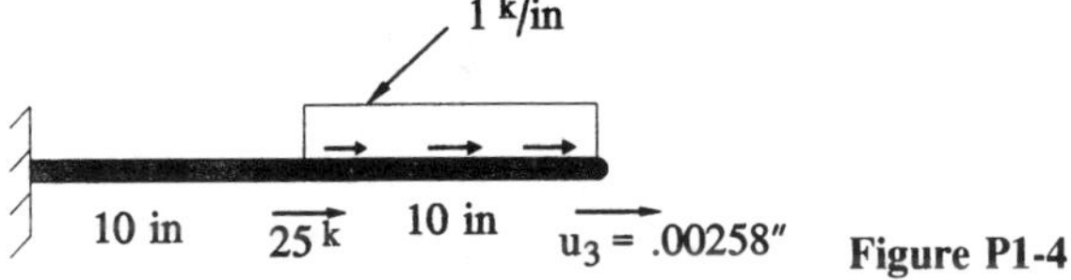

Figure P1-4

1.5 Determine the equivalent nodal forces for the loads shown in Figures P1-5a through P1-5f.

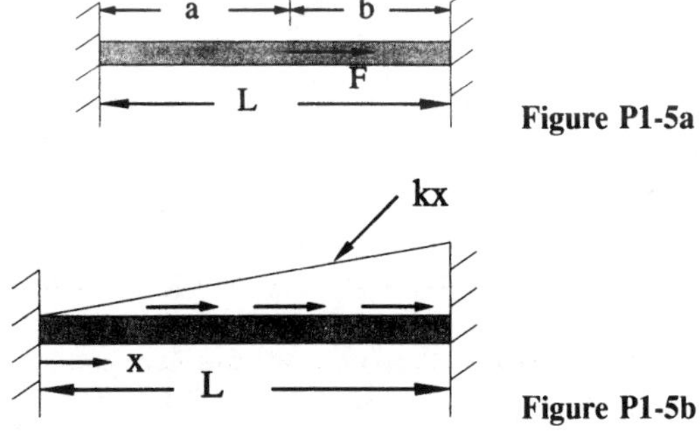

Figure P1-5a

Figure P1-5b

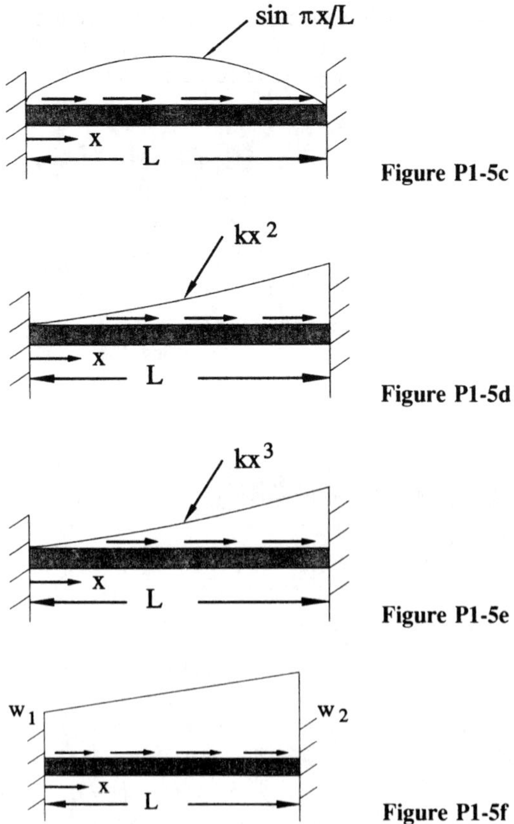

Figure P1-5c

Figure P1-5d

Figure P1-5e

Figure P1-5f

1.6 Using the equivalent nodal forces from problem 1.5, solve for displacements and member forces of the structure shown in Figure P1-6. Show the forces on free body diagrams of each member. $A = 1$ in^2, $E = 10 \times 10^6$ psi.

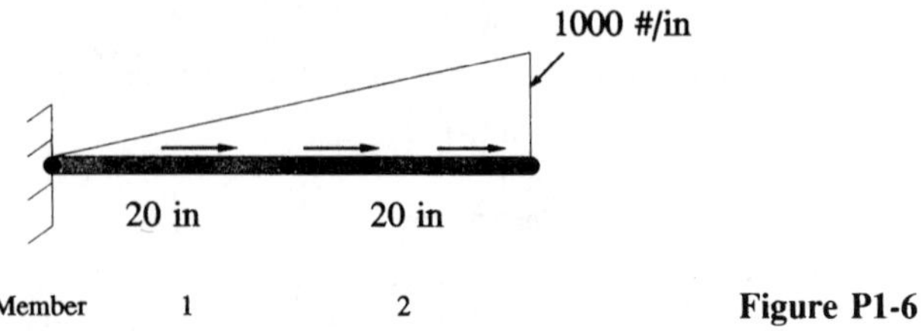

Figure P1-6

1.7 Using the code fragments in this chapter as the basis for your work, write a computer program that will solve one-dimensional bar problems. It should have the capability to use up to 10 elements for modeling a structure. It should calculate nodal displacements and member end forces. You may use the matrix inversion routine in Appendix B if desired. It is not necessary for the program to deal automatically with non-nodal loads or specified non-zero displacements.

1.8 Using the program you wrote in problem 1.7 and the results you obtained from problem 1.5, solve for the nodal displacements and member end forces. Draw free body diagrams of each node and each member showing the directions and magnitudes of all forces acting on them. $A = 1$ in^2, $E = 30 \times 10^6$.

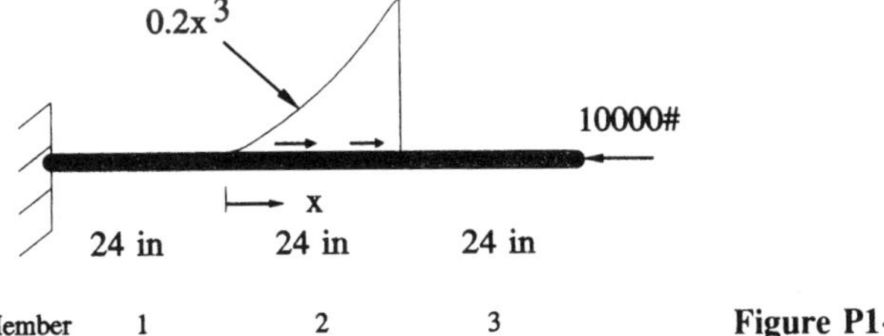

Figure P1-8

1.9 For the structure shown in Figure P1-9, determine the exact solution for the displacement as a function of position along the bar. Model the structure with 2, 4, and 8 elements, and use your computer program to solve for nodal displacements. Compare your computer results with the exact solution. $A = 2$ in^2, $E = 30 \times 10^6$ psi.

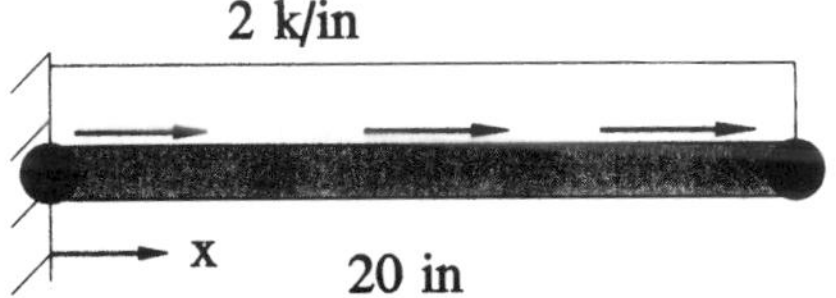

Figure P1-9

1.10 Model the tapered bar shown in Figure P1-10 with 2, 4, and 6 elements, using the average area for each element. Compare your computer results with the exact solution. $E = 29 \times 10^6$ psi.

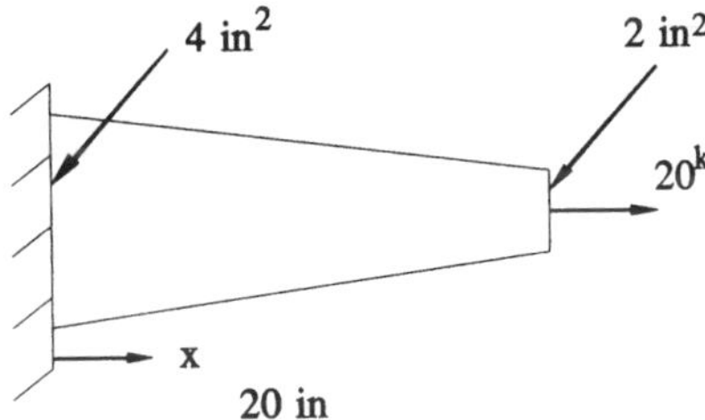

Figure P1-10

1.11 Member 1 in the structure shown in Figure P1-11 undergoes a temperature increase of 50°F. Using $E = 29 \times 10^6$ psi and $\alpha = 6.5 \times 10^{-6}$ in/in/°F, find the nodal displacements and member forces. Draw free-body diagrams of each member.

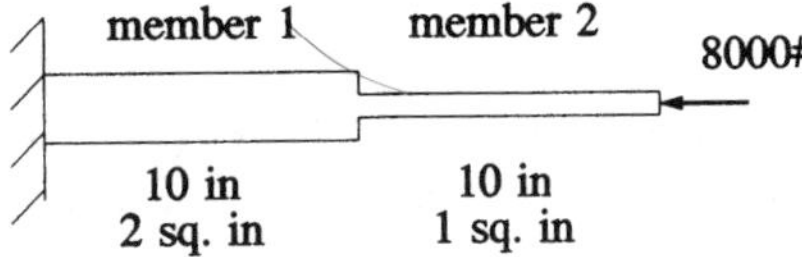

Figure P1-11

1.12 In addition to the distributed load shown acting on the structure depicted in Figure P1-12, member 1 undergoes a temperature change of 30°F. Using $E = 10 \times 10^6$ psi and $\alpha = 12.8 \times 10^{-6}$ in/in/°F, find the nodal displacements and member forces. All areas are 1 in^2.

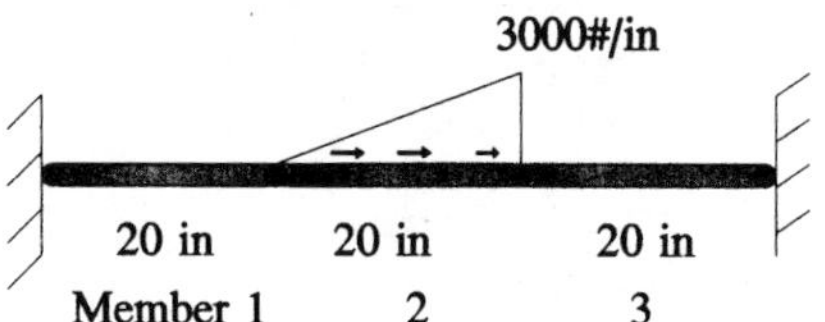

Figure P1-12

1.13 Determine the static and kinematic indeterminacy for the following structures:

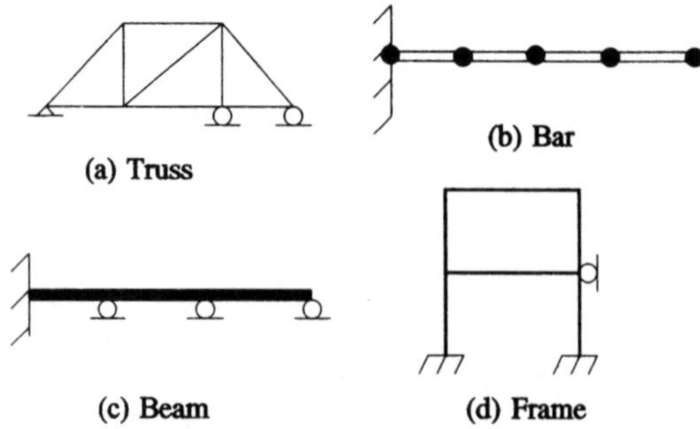

Figure P1-13

CHAPTER 2

ANALYSIS OF TWO-DIMENSIONAL TRUSSES

2.1 INTRODUCTION

You will recall from your basic structures coursework that several assumptions were made when considering the solution of truss problems. They were:

(1) The members of the truss were connected only at their ends.
(2) The connections between members at a joint consisted of frictionless pins.
(3) The members were straight.
(4) Loads were applied only to the joints.

As a consequence of these assumptions we found that each member was a two-force member having only an axial force, either tensile or compressive. Of course, assumptions (2) and (4) above are generally not valid in a real structure such as a timber roof truss, but they do provide us with a "primary" solution to the problem. A "secondary" solution that accounts for end fixity and loads applied intermediate to the joints is then superposed on the primary solution before members are designed.

When we considered the one-dimensional bar element in Chapter 1, the elemental axis was oriented along the length of the bar. The global or system x-axis for the entire structure (the assemblage of bar elements) was also oriented along the axis of the bars. Naturally, there was a single displacement of each node, also oriented along the bar axis. In other words, the elemental x-axis and global x-axis were in the same direction. A two-dimensional truss element, however, can be oriented in any direction in the x-y plane. As a result of this, the elemental axis, which is generally considered to act along the length of the bar from left to right, in general will not be parallel to a global or structural system coordinate axis. In addition, since an unrestrained joint in the structure can displace in the plane of the structure, we must allow for two displacements at each node. Thus, each node will have two degrees of freedom. These displacements are conveniently taken as parallel to the x and y global axes. Since it is always simpler to derive the elemental stiffnesses with respect to the elemental coordinate system, we must find a way to transform the displacements, forces, and elemental stiffness matrices, which are expressed in elemental coordinates, to our global coordinate system before combining to form the structural stiffness matrix. That is, we must have all quantities expressed in terms of a single coordinate system. This requirement leads us to a discussion of vector coordinate transformations.

2.2 COORDINATE TRANSFORMATIONS

Consider the truss member shown in Figure 2-1.

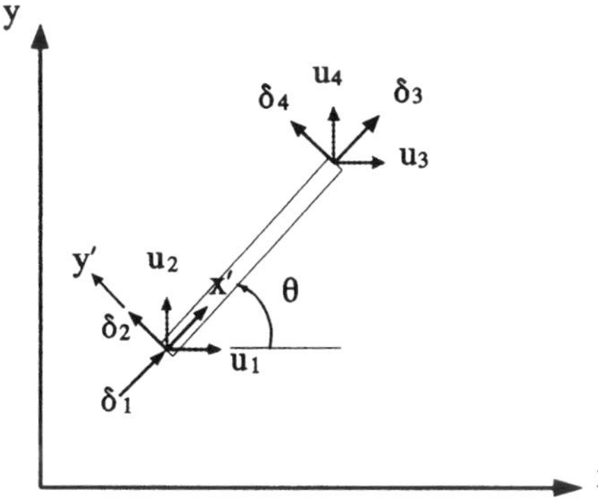

Figure 2-1 Local and global coordinate systems.

Note that the elemental displacements δ_1 through δ_4 are parallel and perpendicular to the member coordinate system x' and y'. Also note the order of numbering of the displacements. Displacements 1 and 2 are at the left end of the member while displacements 3 and 4 are at the right end. We denote the elemental displacements by δ in order to distinguish them from the global displacements u. In order to include the displacements δ_2 and δ_4 in our elemental force-displacement equation, we must expand the elemental force, stiffness, and displacement matrices to a four-degree of freedom system (two degrees of freedom per node). We do this by adding the equations $P_2 = 0$ and $P_4 = 0$. Our elemental force-displacement relationship $\{P\} = [k]\{\delta\}$ becomes

$$\begin{Bmatrix} P_1 \\ P_2 \\ P_3 \\ P_4 \end{Bmatrix} = EA/L \begin{bmatrix} 1 & 0 & -1 & 0 \\ 0 & 0 & 0 & 0 \\ -1 & 0 & 1 & 0 \\ 0 & 0 & 0 & 0 \end{bmatrix} \begin{Bmatrix} \delta_1 \\ \delta_2 \\ \delta_3 \\ \delta_4 \end{Bmatrix} \tag{2.1}$$

Note that the elemental forces which act in the elemental coordinate directions x'–y' are denoted by P in order to distinguish them from the global forces F, which act parallel to the global x-y coordinate system. We would like to express the elemental displacements in terms of the global or system displacements. This transformation will take the form

$$\{\delta\} = [\beta]\{u\} \tag{2.2}$$

Consider the vector displacement of the left end of the member as shown in Figure 2-2.

In Figure 2-2, θ is the angle between the global x-axis and the element x-axis, measured positive counterclockwise. We can see that

$$\delta_1 = u_1 \cos\theta + u_2 \sin\theta$$

$$\delta_2 = u_2 \cos\theta - u_1 \sin\theta$$

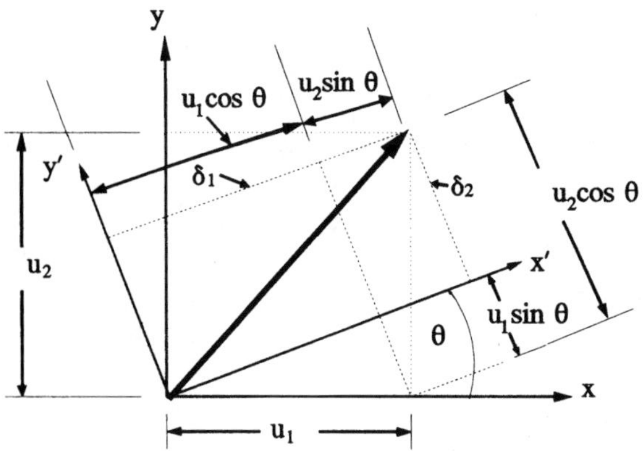

Figure 2-2 Components of displacement vector.

The same relationships between displacements will also exist at the right end of the member. Thus

$$\delta_3 = u_3 \cos\theta + u_4 \sin\theta$$

$$\delta_4 = u_4 \cos\theta - u_3 \sin\theta$$

In matrix form, equation (2.2), $\{\delta\} = [\beta]\{u\}$, becomes

$$\begin{Bmatrix} \delta_1 \\ \delta_2 \\ \delta_3 \\ \delta_4 \end{Bmatrix} = \begin{bmatrix} \cos\theta & \sin\theta & 0 & 0 \\ -\sin\theta & \cos\theta & 0 & 0 \\ 0 & 0 & \cos\theta & \sin\theta \\ 0 & 0 & -\sin\theta & \cos\theta \end{bmatrix} \begin{Bmatrix} u_1 \\ u_2 \\ u_3 \\ u_4 \end{Bmatrix} \tag{2.3}$$

Now, the length of the displacement vector must be the same in both the system and the elemental coordinate systems.

$$\delta_1^2 + \delta_2^2 = u_1^2 + u_2^2$$

$$[\,\delta_1 \quad \delta_2\,] \begin{Bmatrix} \delta_1 \\ \delta_2 \end{Bmatrix} = [\,u_1 \quad u_2\,] \begin{Bmatrix} u_1 \\ u_2 \end{Bmatrix}$$

which can be written

$$[\delta]^T \{\delta\} = [u]^T \{u\} \tag{2.4}$$

Since $\{\delta\} = [\beta]\{u\}$ then

$$\{u\} = [\beta]^{-1} \{\delta\} \tag{2.5}$$

From matrix algebra, if $[A] = [B][C]$, then $[A]^T = [C]^T[B]^T$. Thus,

$$\{\delta\}^T = [u]^T [\beta]^T \tag{2.6}$$

Using equations (2.5) and (2.6) in equation (2.4) yields

$$[u]^T [\beta]^T \{\delta\} = [u]^T [\beta]^{-1} \{\delta\}$$

Thus,

$$[\beta]^T = [\beta]^{-1} \tag{2.7}$$

Equation (2.7) is a property of rotation of an orthogonal coordinate system.

2.3 GLOBAL STIFFNESS MATRIX

When we combined the elemental stiffness matrices for the one-dimensional rod in order to generate the system or global stiffness matrix, it was done directly by identifying stiffness terms with nodal displacements. This was possible since all elemental stiffnesses were also stiffnesses with respect to the global coordinate system (remember that the coordinate systems were identical). This is not the case for the truss element. The local (elemental) and global coordinate systems are not, in general, the same. Before we can combine the elemental stiffnesses in order to generate the global stiffness matrix, we must transform each elemental stiffness to a common set of axes. We naturally choose the global set of axes as this common set.

Equation (2.3), $\{\delta\} = [\beta]\{u\}$, is the transformation equation for a two-dimensional vector. It can therefore be used for forces as well as displacements. That is, we can write

$$\{P\} = [\beta]\{F\} \tag{2.8}$$

where $\{P\}$ represents the elemental forces with respect to the elemental coordinate system and $\{F\}$ represents the elemental forces with respect to the global coordinate system.

For the elemental coordinate system we have

$$\{P\} = [k]\{\delta\} \tag{2.9}$$

Using equation (2.9) in equation (2.8) we find

$$[k]\{\delta\} = [\beta]\{F\} \tag{2.10}$$

Substituting equation (2.2) into (2.10) yields

$$[k][\beta]\{u\} = [\beta]\{F\} \tag{2.11}$$

Solving for $\{F\}$ we have

$$\{F\} = [\beta]^{-1}[k][\beta]\{u\} \tag{2.12}$$

Since the force-displacement relationship for the global coordinate system can be written $\{F\} = [K]\{u\}$, then from equation (2.12) we see that

$$[K] = [\beta]^{-1}[k][\beta]$$

which by virtue of equation (2.7) becomes

$$[K] = [\beta]^T[k][\beta] \tag{2.13}$$

Equation (2.13) is used to transform the elemental stiffness matrix $[k]$ with respect to the elemental coordinate system, to the elemental stiffness matrix $[K]$ with respect to the global coordinate system. Once the stiffnesses of each element in our structure have been transformed in accordance with equation (2.13), we will be able to combine them into a global structural stiffness matrix. We can then write for the entire structure

$$\{F\}_{global} = [K]_{global}\,\{u\}_{global}$$

where the "global" subscript refers to the entire structure.

Expanding equation (2.13) and designating the sine and cosine terms by S and C, respectively, we have

$$[K]_{system\ coord.} = EA/L \begin{bmatrix} C & -S & 0 & 0 \\ S & C & 0 & 0 \\ 0 & 0 & C & -S \\ 0 & 0 & S & C \end{bmatrix} \begin{bmatrix} 1 & 0 & -1 & 0 \\ 0 & 0 & 0 & 0 \\ -1 & 0 & 1 & 0 \\ 0 & 0 & 0 & 0 \end{bmatrix} \begin{bmatrix} C & S & 0 & 0 \\ -S & C & 0 & 0 \\ 0 & 0 & C & S \\ 0 & 0 & -S & C \end{bmatrix}$$

$$= EA/L \begin{bmatrix} C^2 & CS & -C^2 & -CS \\ CS & S^2 & -CS & -S^2 \\ -C^2 & -CS & C^2 & CS \\ -CS & -S^2 & CS & S^2 \end{bmatrix} \tag{2.14}$$

Equation (2.14) is the elemental stiffness matrix in terms of system (global) coordinates. Thus, we can formulate the system stiffness matrix by summing the elemental stiffness matrices after they have been transformed to the global coordinate system.

2.3.1 Truss Examples

Example 2.1

Consider the truss shown in Figure E2-1.

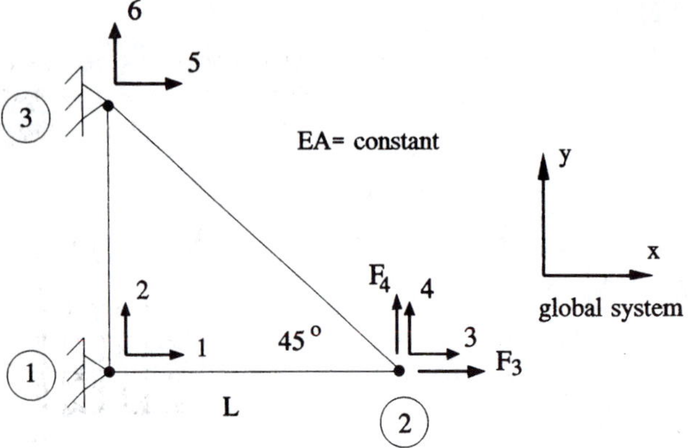

Figure E2-1

The system displacements with respect to the global coordinate system are as follows:

Node	System displacements
1	u_1, u_2
2	u_3, u_4
3	u_5, u_6

Note that the x displacement at each node has a subscript equal to twice the node number minus 1, and the y displacement subscript has a subscript equal to twice the node number. We will use this scheme to designate displacements for a truss.

For element number 1, consider nodes number 1 and 2 as the left and right ends of the member, respectively. Thus, for element number 1, with $\theta_{1-2} = 0°$ and $C = 1$, $S = 0$, we have

$$\begin{matrix} & & 1 & 2 & 3 & 4 & & \\ \begin{Bmatrix} F_1 \\ F_2 \\ F_3 \\ F_4 \end{Bmatrix} = EA/L & & \left[\begin{matrix} 1 \\ 0 \\ -1 \\ 0 \end{matrix}\right. & \begin{matrix} 0 \\ 0 \\ 0 \\ 0 \end{matrix} & \begin{matrix} -1 \\ 0 \\ 1 \\ 0 \end{matrix} & \left.\begin{matrix} 0 \\ 0 \\ 0 \\ 0 \end{matrix}\right] & \begin{Bmatrix} u_1 \\ u_2 \\ u_3 \\ u_4 \end{Bmatrix} & \begin{matrix} 1 \\ 2 \\ 3 \\ 4 \end{matrix} \end{matrix}$$

For element number 2, nodes 2 and 3 locate the left and right ends of the member. Therefore $\theta_{2-3} = 135°$, and $C = -1/\sqrt{2}$, $S = 1/\sqrt{2}$. Thus,

$$\begin{matrix} & & 3 & 4 & 5 & 6 & & \\ \begin{Bmatrix} F_3 \\ F_4 \\ F_5 \\ F_6 \end{Bmatrix} = EA/\sqrt{2}L & & \left[\begin{matrix} 1/2 \\ -1/2 \\ -1/2 \\ 1/2 \end{matrix}\right. & \begin{matrix} -1/2 \\ 1/2 \\ 1/2 \\ -1/2 \end{matrix} & \begin{matrix} -1/2 \\ 1/2 \\ 1/2 \\ -1/2 \end{matrix} & \left.\begin{matrix} 1/2 \\ -1/2 \\ -1/2 \\ 1/2 \end{matrix}\right] & \begin{Bmatrix} u_3 \\ u_4 \\ u_5 \\ u_6 \end{Bmatrix} & \begin{matrix} 3 \\ 4 \\ 5 \\ 6 \end{matrix} \end{matrix}$$

For element 3, nodes 1 and 3 are located at the left and right ends of the member, so $\theta_{1-3} = 90°$ and $S = 1$, $C = 0$.

$$\begin{matrix} & & 1 & 2 & 5 & 6 & & \\ \begin{Bmatrix} F_1 \\ F_2 \\ F_5 \\ F_6 \end{Bmatrix} = EA/L & & \left[\begin{matrix} 0 \\ 0 \\ 0 \\ 0 \end{matrix}\right. & \begin{matrix} 0 \\ 1 \\ 0 \\ -1 \end{matrix} & \begin{matrix} 0 \\ 0 \\ 0 \\ 0 \end{matrix} & \left.\begin{matrix} 0 \\ -1 \\ 0 \\ 1 \end{matrix}\right] & \begin{Bmatrix} u_1 \\ u_2 \\ u_5 \\ u_6 \end{Bmatrix} & \begin{matrix} 1 \\ 2 \\ 5 \\ 6 \end{matrix} \end{matrix}$$

Since we now have all elemental stiffnesses expressed in terms of the global coordinate system, we can now construct the system stiffness matrix. The structure has three nodes and therefore six degrees of freedom. The structural stiffness matrix will be a 6×6 matrix. Accumulating elements of the elemental stiffness matrices using the global codes noted above and to the right of the matrices we find

$$\begin{Bmatrix} F_1 \\ F_2 \\ F_3 \\ F_4 \\ F_5 \\ F_6 \end{Bmatrix} = EA/L \begin{bmatrix} 1 & 0 & -1 & 0 & 0 & 0 \\ 0 & 1 & 0 & 0 & 0 & -1 \\ -1 & 0 & 1+1/2\sqrt{2} & -1/2\sqrt{2} & -1/2\sqrt{2} & 1/2\sqrt{2} \\ 0 & 0 & -1/2\sqrt{2} & 1/2\sqrt{2} & 1/2\sqrt{2} & -1/2\sqrt{2} \\ 0 & 0 & -1/2\sqrt{2} & 1/2\sqrt{2} & 1/2\sqrt{2} & -1/2\sqrt{2} \\ 0 & -1 & 1/2\sqrt{2} & -1/2\sqrt{2} & -1/2\sqrt{2} & 1+1/2\sqrt{2} \end{bmatrix} \begin{Bmatrix} u_1 \\ u_2 \\ u_3 \\ u_4 \\ u_5 \\ u_6 \end{Bmatrix} \tag{2.15}$$

where each term in the force matrix represents the total nodal force in a global x or y coordinate direction at a node. These forces are either reactions or applied forces. As an example of the combination of stiffnesses, note that a row 6 and column 6 designation appears in the elemental stiffness matrices for both elements number 2 and 3. Thus, element 6,6 in the structural stiffness matrix is the sum of these two individual stiffnesses.

We now reduce equation (2.15) by applying support condition constraints. Both nodes 1 and 3 are pinned. As a result, displacements u_1, u_2, u_5, and u_6 are zero. Eliminating the rows and columns associated with these zero displacements results in the reduced stiffness matrix shown in equation (2.16).

$$\begin{Bmatrix} F_3 \\ F_4 \end{Bmatrix} = EA/L \begin{bmatrix} 1 + 1/2\sqrt{2} & -1/2\sqrt{2} \\ -1/2\sqrt{2} & 1/2\sqrt{2} \end{bmatrix} \begin{Bmatrix} u_3 \\ u_4 \end{Bmatrix} \tag{2.16}$$

Solving equation (2.16) for the two unknown displacements u_3 and u_4 yields

$$u_3 = (F_3 + F_4)L/EA$$

$$u_4 = \left[(1 + 2\sqrt{2})F_4 + F_3\right] L/EA$$

The reactions F_1, F_2, F_5, and F_6 are found by substituting these displacements into equation (2.15). We find,

$$F_1 = -(F_3 + F_4), \quad F_2 = 0, \quad F_5 = F_4, \quad \text{and } F_6 = -F_4.$$

Sketching these reactions and the applied loads F_3 and F_4 on the structure as shown below, we verify overall equilibrium.

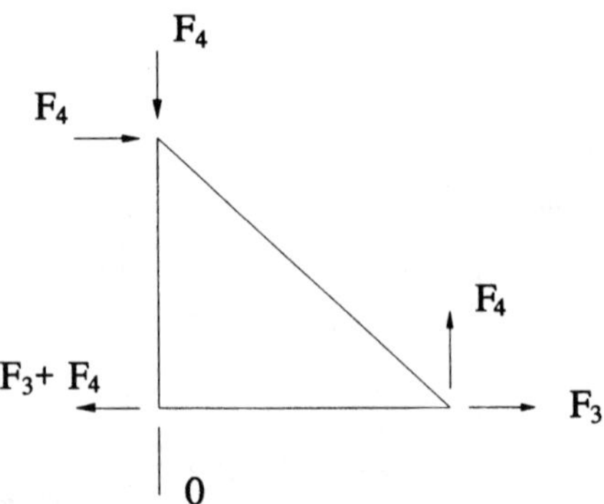

We next calculate the member forces.

Since $\{\delta\} = [\beta]\{u\}$ and $\{P\} = [k]\{\delta\}$, then $\{P\} = [k][\beta]\{u\}$, where $[k]$ is the elemental stiffness matrix with respect to the elemental coordinate system. The $[\beta]$ and $\{u\}$ matrices are different for each element.

For element number 1,

$$\begin{Bmatrix} P_1 \\ P_2 \\ P_3 \\ P_4 \end{Bmatrix} = EA/L \begin{bmatrix} 1 & 0 & -1 & 0 \\ 0 & 0 & 0 & 0 \\ -1 & 0 & 1 & 0 \\ 0 & 0 & 0 & 0 \end{bmatrix} \begin{bmatrix} 1 & 0 & 0 & 0 \\ 0 & 1 & 0 & 0 \\ 0 & 0 & 1 & 0 \\ 0 & 0 & 0 & 1 \end{bmatrix} \begin{Bmatrix} u_1 \\ u_2 \\ u_3 \\ u_4 \end{Bmatrix} = \begin{Bmatrix} -(F_3 + F_4) \\ 0 \\ (F_3 + F_4) \\ 0 \end{Bmatrix}$$

For element number 2,

$$\begin{Bmatrix} P_1 \\ P_2 \\ P_3 \\ P_4 \end{Bmatrix} = EA/\sqrt{2}L \begin{bmatrix} 1 & 0 & -1 & 0 \\ 0 & 0 & 0 & 0 \\ -1 & 0 & 1 & 0 \\ 0 & 0 & 0 & 0 \end{bmatrix} \begin{bmatrix} -1/\sqrt{2} & 1/\sqrt{2} & 0 & 0 \\ -1/\sqrt{2} & -1/\sqrt{2} & 0 & 0 \\ 0 & 0 & -1/\sqrt{2} & 1/\sqrt{2} \\ 0 & 0 & -1/\sqrt{2} & -1/\sqrt{2} \end{bmatrix} \begin{Bmatrix} u_3 \\ u_4 \\ u_5 \\ u_6 \end{Bmatrix}$$

$$= 1/\sqrt{2} \begin{Bmatrix} 2F_4 \\ 0 \\ -2F_4 \\ 0 \end{Bmatrix}$$

and for element number 3,

$$\begin{Bmatrix} P_1 \\ P_2 \\ P_3 \\ P_4 \end{Bmatrix} = EA/L \begin{bmatrix} 1 & 0 & -1 & 0 \\ 0 & 0 & 0 & 0 \\ -1 & 0 & 1 & 0 \\ 0 & 0 & 0 & 0 \end{bmatrix} \begin{bmatrix} 0 & 1 & 0 & 0 \\ -1 & 0 & 0 & 0 \\ 0 & 0 & 0 & 1 \\ 0 & 0 & -1 & 0 \end{bmatrix} \begin{Bmatrix} u_1 \\ u_2 \\ u_5 \\ u_6 \end{Bmatrix} = \begin{Bmatrix} 0 \\ 0 \\ 0 \\ 0 \end{Bmatrix}$$

Note that we need to calculate only P_3 for each member since it represents the axial force in the member, positive when tensile.

Example 2.2

Consider the truss shown in Figure E2-2. EA = constant.

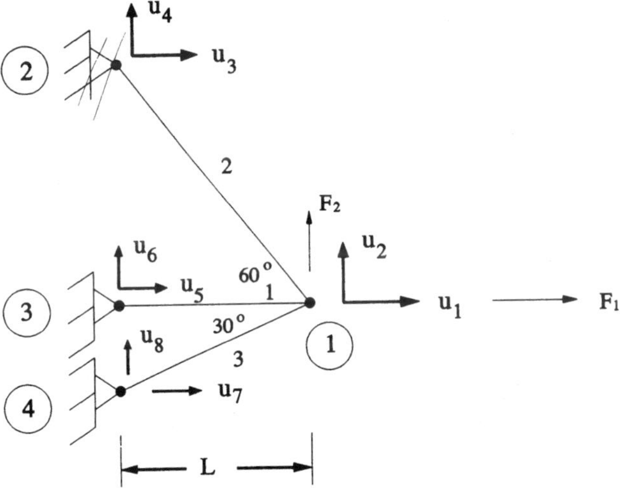

Figure E2-2

Let node number 1 be the left end of each member. From equation (2.14) the transformed elemental stiffness matrices become:

Member 1:

$\theta_{1-3} = 180°$, $L_{1-3} = L$

$$[k]_1 = EA/L \begin{bmatrix} 1 & 0 & -1 & 0 \\ 0 & 0 & 0 & 0 \\ -1 & 0 & 1 & 0 \\ 0 & 0 & 0 & 0 \end{bmatrix}$$

Member 2:

$\theta_{1-2} = 120°$, $L_{1-2} = 2L$

$$[k]_2 = EA/L \begin{bmatrix} .125 & -.2165 & -.125 & .2165 \\ -.2165 & .375 & .2165 & -.375 \\ -.125 & .2165 & .125 & -.2165 \\ .2165 & -.375 & -.2165 & .375 \end{bmatrix}$$

Member 3:

$\theta_{1-4} = 210°$, $L_{1-4} = 1.155L$

$$[k]_3 = EA/L \begin{bmatrix} .65 & .375 & -.65 & -.375 \\ .375 & .2165 & -.375 & -.2165 \\ -.65 & -.375 & .65 & .375 \\ -.375 & -.2165 & .375 & .2165 \end{bmatrix}$$

Now,

$$\begin{Bmatrix} F_1 \\ F_2 \\ F_5 \\ F_6 \end{Bmatrix} = [k]_1 \begin{Bmatrix} u_1 \\ u_2 \\ u_5 \\ u_6 \end{Bmatrix} \begin{Bmatrix} F_1 \\ F_2 \\ F_3 \\ F_4 \end{Bmatrix} = [k]_2 \begin{Bmatrix} u_1 \\ u_2 \\ u_3 \\ u_4 \end{Bmatrix} \begin{Bmatrix} F_1 \\ F_2 \\ F_7 \\ F_8 \end{Bmatrix} = [k]_3 \begin{Bmatrix} u_1 \\ u_2 \\ u_7 \\ u_8 \end{Bmatrix}$$

Combining, we find

$$\begin{Bmatrix} F_1 \\ F_2 \\ F_3 \\ F_4 \\ F_5 \\ F_6 \\ F_7 \\ F_8 \end{Bmatrix} = EA/L \begin{bmatrix} 1.775 & .1585 & -.125 & .2165 & -1 & 0 & -.65 & -.375 \\ .1585 & .5915 & .2165 & -.375 & 0 & 0 & -.375 & -.2165 \\ -.125 & .2165 & .125 & -.2165 & 0 & 0 & 0 & 0 \\ .2165 & -.375 & -.2165 & .375 & 0 & 0 & 0 & 0 \\ -1 & 0 & 0 & 0 & 1 & 0 & 0 & 0 \\ 0 & 0 & 0 & 0 & 0 & 0 & 0 & 0 \\ -.65 & -.375 & 0 & 0 & 0 & 0 & .65 & .375 \\ -.375 & -.2165 & 0 & 0 & 0 & 0 & .375 & .2165 \end{bmatrix} \begin{Bmatrix} u_1 \\ u_2 \\ u_3 \\ u_4 \\ u_5 \\ u_6 \\ u_7 \\ u_8 \end{Bmatrix} \tag{2.17}$$

Noting that only u_1 and u_2 are non-zero, our reduced equation becomes

$$\begin{Bmatrix} F_1 \\ F_2 \end{Bmatrix} = EA/L \begin{bmatrix} 1.775 & .1585 \\ .1585 & .5915 \end{bmatrix} \begin{Bmatrix} u_1 \\ u_2 \end{Bmatrix} \tag{2.18}$$

Solving for the displacements we find

$$\begin{Bmatrix} u_1 \\ u_2 \end{Bmatrix} = L/EA \begin{bmatrix} .577 & -.1547 \\ -.1547 & 1.732 \end{bmatrix} \begin{Bmatrix} F_1 \\ F_2 \end{Bmatrix} \tag{2.19}$$

Thus,

$$u_1 = (L/EA)(.577F_1 - .1547F_2)\ u_2 = (L/EA)(-.1547F_1 + 1.732F_2)$$

Using equation (2.17), the reactions are found to be

$$\begin{Bmatrix} F_3 \\ F_4 \\ F_5 \\ F_6 \\ F_7 \\ F_8 \end{Bmatrix} = EA/L \begin{bmatrix} -.125 & .2165 \\ .2165 & -.375 \\ -1 & 0 \\ 0 & 0 \\ -.65 & -.375 \\ -.375 & -.2165 \end{bmatrix} \begin{Bmatrix} u_1 \\ u_2 \end{Bmatrix} \tag{2.20}$$

Using $\{P\} = [k][\beta]\{u\}$ and calculating P_3 for each member gives

$$P_{member\ 1} = (.577F_1 - .1547F_2)$$

$$P_{member\ 2} = (.211F_1 - .789F_2)$$

$$P_{member\ 3} = (.365F_1 + .634F_2)$$

A free-body diagram of node 1 appears below.

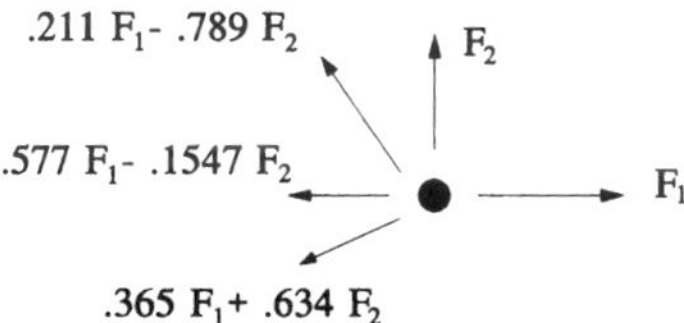

Resolving the forces into the horizontal and vertical directions, and writing the equilibrium equations yields

$$\Sigma F_x = -.1055\ F_1 - .577\ F_1 - .3161\ F_1 + F_1 + .3945\ F_2 + .1547\ F_2 - .549\ F_2 \approx 0$$

$$\Sigma F_y = F_1(.183 - .183) + F_2(-.683 - .317 + 1) = 0$$

2.4 SUPPORT MOVEMENTS

We often need to account for specified support displacements. These displacements could be due to soil consolidation or expansion or to non-precise placement of foundations of the structure.

In the first case, the geotechnical engineer will make an estimate of the possible movement of supports based upon soil profiles, moisture content and its variation, and both dead and live loads transferred to the foundations by the structure.

In the case of imprecise foundation placement, which is generally discovered during erection of the structure, a survey is performed to determine the magnitudes of the discrepancies between the specified position and actual location. In either case, the support displacements will be known.

Techniques for dealing with prescribed nodal displacements for the one-dimensional rod were presented in Chapter 1. Of course, these techniques can be used for any structure with support movements, and both techniques will be illustrated for a two-dimensional truss.

As in section 1.5, suppose we write the structural equation in the following way:

$$\left\{\begin{matrix}\{F_p\}\\\{F_s\}\end{matrix}\right\} = \begin{bmatrix}[K_{pp}] & [K_{ps}]\\ [K_{sp}] & [K_{ss}]\end{bmatrix}\left\{\begin{matrix}\{u_p\}\\\{u_s\}\end{matrix}\right\} \tag{2.21}$$

where $\{F_p\}$ and $\{u_p\}$ represent known applied nodal forces and corresponding nodal displacements, and $\{F_s\}$ and $\{u_s\}$ represent unknown support reactions and corresponding support displacements (some of which may be zero). Note that this requires reordering of the rows and columns of the original matrix equation. Expanding equation (2.21) yields

$$\{F_p\} = [K_{pp}]\{u_p\} + [K_{ps}]\{u_s\} \tag{2.22}$$

$$\{F_s\} = [K_{sp}]\{u_p\} + [K_{ss}]\{u_s\} \tag{2.23}$$

Note that if $\{u_s\} = \{0\}$, that is, if all support displacements are zero, the above equations become

$$\{F_p\} = [K_{pp}]\{u_p\} \tag{2.24}$$

$$\{F_s\} = [K_{sp}]\{u_p\} \tag{2.25}$$

The terms in equation (2.24) are identical to the reduced force, stiffness, and displacement matrices obtained by removing the rows and columns associated with the zero displacements.

Once the displacements $\{u_p\}$ have been found by solving equation (2.24), equation (2.25) is used to determine the reactions.

Consider the case where some or all terms of $\{u_s\}$ are non-zero. Solving equation (2.22) for $\{u_p\}$ yields

$$\{u_p\} = [K_{pp}]^{-1}\left(\{F_p\} - [K_{ps}]\{u_s\}\right) \tag{2.26}$$

We now use the values of $\{u_p\}$ in equation (2.23) to determine the forces associated with the specified displacements.

Keep in mind that for a determinate structure, no additional bar forces will be introduced by support movements. Since a determinate structure is not overconstrained, it is able to accommodate support movements by altering its configuration through displacements of the nodes. That is, the members undergo rigid body motion and therefore develop no additional stresses.

Let us solve the following problem using both techniques presented.

Example 2.3

Use the technique presented in Chapter 1 to solve for nodal displacements, reactions, and member forces of the truss shown in Figure E2-3. The support at node 1 displaces down 0.6 in and node 4 displaces to the left 0.3 in. All areas are 2 in^2 and $E = 29 \times 10^6$ psi.

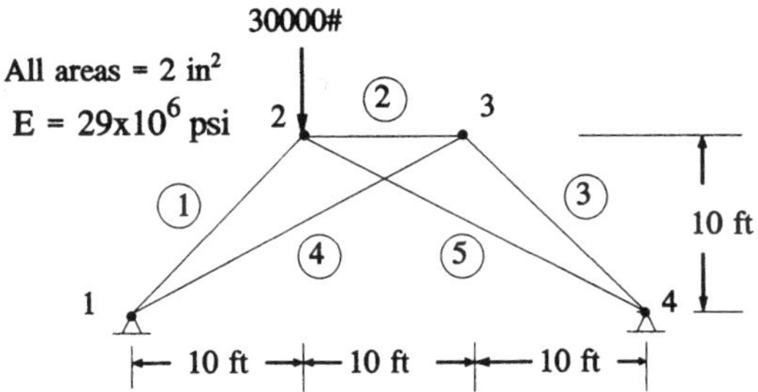

Figure E2-3

Step 1: Generate the overall structural stiffness equation. This equation is shown below.

$$\begin{Bmatrix} F_1 \\ F_2 \\ F_3 \\ F_4 \\ F_5 \\ F_6 \\ F_7 \\ F_8 \end{Bmatrix} = \begin{bmatrix} 343806 & 257345 & -170884 & -170884 & -172923 & -86461 & 0 & 0 \\ 257345 & 214115 & -170884 & -170884 & -86461 & -43231 & 0 & 0 \\ -170884 & -170884 & 827140 & 84423 & -483333 & 0 & -172923 & 86461 \\ -170884 & -170884 & 84423 & 214115 & 0 & 0 & 86461 & -43231 \\ -172923 & -86461 & -483333 & 0 & 827140 & -84423 & -170884 & 170884 \\ -86461 & -43231 & 0 & 0 & -84423 & 214115 & 170884 & -170884 \\ 0 & 0 & -172923 & 86461 & -170884 & 170884 & 343807 & -257345 \\ 0 & 0 & 86461 & -43231 & 170884 & -170884 & -257345 & 214115 \end{bmatrix} \begin{Bmatrix} u_1 \\ u_2 \\ u_3 \\ u_4 \\ u_5 \\ u_6 \\ u_7 \\ u_8 \end{Bmatrix} \tag{2.27}$$

Step 2: Eliminate the rows and columns associated with zero displacements. In this case, u_1 and u_8 are zero. We therefore eliminate rows 1 and 8, and columns 1 and 8.

Step 3: If a displacement u_k is specified at coordinate n, multiply k_{nn} by a large number M and replace the force value in row n by $u_k \times M \times k_{nn}$. In this example, displacements u_2 and u_7 are specified. We therefore multiply k_{22} and k_{77} by a large number (taken as 10^9), and replace the force values in rows 2 and 7 by the appropriate products. Equation (2.28) is the result of these manipulations.

$$\begin{Bmatrix} -1.2847 \times 10^{14} \\ 0 \\ -30000 \\ 0 \\ 0 \\ -1.0314 \times 10^{14} \end{Bmatrix} = \begin{bmatrix} 214115 \times 10^9 & -170884 & -170884 & -86461 & -43231 & 0 \\ -170884 & 827140 & 84423 & -483333 & 0 & -172923 \\ -170884 & 84423 & 214115 & 0 & 0 & 86461 \\ -86461 & -483333 & 0 & 827140 & -84423 & -170884 \\ -43231 & 0 & 0 & -84423 & 214115 & 170884 \\ 0 & -172923 & 86461 & -170884 & 170884 & 343807 \times 10^9 \end{bmatrix} \begin{Bmatrix} u_2 \\ u_3 \\ u_4 \\ u_5 \\ u_6 \\ u_7 \end{Bmatrix} \tag{2.28}$$

Solving equation (2.28) for the displacements, we find

$$\begin{Bmatrix} u_2 \\ u_3 \\ u_4 \\ u_5 \\ u_6 \\ u_7 \end{Bmatrix} = \begin{Bmatrix} -.60000 \\ -.33848 \\ -.36437 \\ -.32343 \\ -.00925 \\ -.30000 \end{Bmatrix} \text{ inches} \tag{2.29}$$

The reactions are found from equation (2.27) by expanding rows 1, 2, 7, and 8. The results are

$$\begin{Bmatrix} F_1 \\ F_2 \\ F_7 \\ F_8 \end{Bmatrix} = \begin{Bmatrix} 22426 \\ 20001 \\ -22427 \\ 10001 \end{Bmatrix} \text{ pounds} \tag{2.30}$$

Note from the sketch shown below that overall equilibrium checks with very little round-off error.

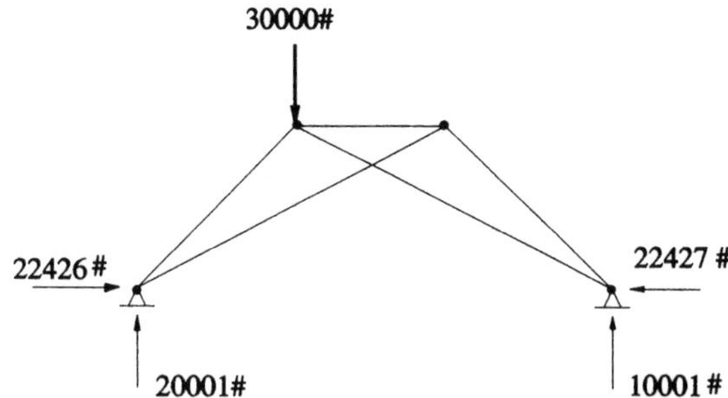

The member forces are now obtained from $\{P\} = [k][\beta]\{u\}$. As an example, consider member number 1. With $EA/L = 341{,}768.3$#/in and $\theta = 45°$, we have

$$\begin{Bmatrix} P_1 \\ P_2 \\ P_3 \\ P_4 \end{Bmatrix} = 341768.3 \begin{bmatrix} 1 & 0 & -1 & 0 \\ 0 & 0 & 0 & 0 \\ -1 & 0 & 1 & 0 \\ 0 & 0 & 0 & 0 \end{bmatrix} \begin{bmatrix} .7071 & .7071 & 0 & 0 \\ -.7071 & .7071 & 0 & 0 \\ 0 & 0 & .7071 & .7071 \\ 0 & 0 & -.7071 & .7071 \end{bmatrix} \begin{Bmatrix} u_1 \\ u_2 \\ u_3 \\ u_4 \end{Bmatrix} \tag{2.31}$$

which gives

$$\begin{Bmatrix} P_1 \\ P_2 \\ P_3 \\ P_4 \end{Bmatrix} = 341768.3 \begin{Bmatrix} .0727246 \\ 0 \\ -.0727246 \\ 0 \end{Bmatrix} = \begin{Bmatrix} 24855 \\ 0 \\ -24855 \\ 0 \end{Bmatrix} \text{pounds} \tag{2.32}$$

Similarly, for member number 4, with $EA/L = 216{,}153.2$#/in,

$$\begin{Bmatrix} P_1 \\ P_2 \\ P_3 \\ P_4 \end{Bmatrix} = 216153.2 \begin{bmatrix} 1 & 0 & -1 & 0 \\ 0 & 0 & 0 & 0 \\ -1 & 0 & 1 & 0 \\ 0 & 0 & 0 & 0 \end{bmatrix} \begin{bmatrix} .8944 & .4472 & 0 & 0 \\ -.4472 & .8944 & 0 & 0 \\ 0 & 0 & .8944 & .4472 \\ 0 & 0 & -.4472 & .8944 \end{bmatrix} \begin{Bmatrix} u_1 \\ u_2 \\ u_5 \\ u_6 \end{Bmatrix}$$

which yields

$$\begin{Bmatrix} P_1 \\ P_2 \\ P_3 \\ P_4 \end{Bmatrix} = \begin{Bmatrix} 5425 \\ 0 \\ -5425 \\ 0 \end{Bmatrix} \text{pounds}$$

Sketching the forces at node 1 we have

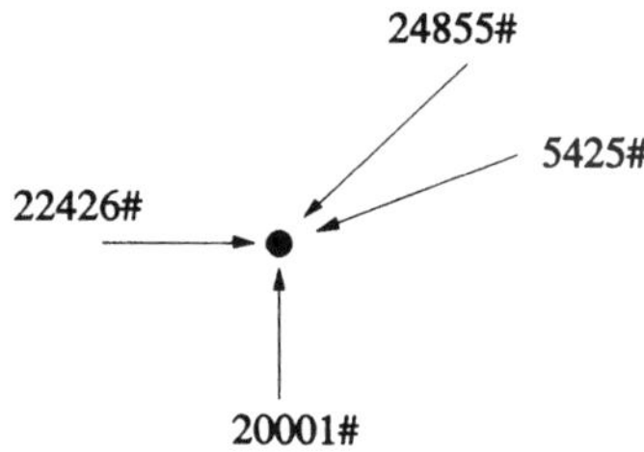

Resolving these forces into x and y components provides a joint equilibrium check.

These checks are very important to perform, particularly when computer programs are used for solving problems. They enable errors in input data and/or the program itself to be detected.

Example 2.4

Solve the previous problem using the matrix partitioning technique.

Since u_1, u_2, u_7, and u_8 are the known displacements, we must reorder the original structural stiffness matrix in the following way:

$$\begin{Bmatrix} F_3 \\ F_4 \\ F_5 \\ F_6 \\ F_1 \\ F_2 \\ F_7 \\ F_8 \end{Bmatrix} = \begin{bmatrix} 827140 & 84423 & -483333 & 0 & -170884 & -170884 & -172923 & 86461 \\ 84423 & 214115 & 0 & 0 & -170884 & -170884 & 86461 & -43231 \\ -483333 & 0 & 827140 & -84423 & -172923 & -86461 & -170884 & 170884 \\ 0 & 0 & -84423 & 214115 & -86461 & -43231 & 170884 & -170884 \\ -170884 & -170884 & -172923 & -86461 & 343807 & 257345 & 0 & 0 \\ -170884 & -170884 & -86461 & -43231 & 257345 & 214115 & 0 & 0 \\ -172923 & 86461 & -170884 & 170884 & 0 & 0 & 343807 & -257345 \\ 86461 & -43231 & 170884 & -170884 & 0 & 0 & -257345 & 214115 \end{bmatrix} \begin{Bmatrix} u_3 \\ u_4 \\ u_5 \\ u_6 \\ u_1 \\ u_2 \\ u_7 \\ u_8 \end{Bmatrix} \tag{2.33}$$

In this example, each of the submatrices is a 4×4 matrix. This will not always be the case. Using equation (2.26) we have

$$[K_{pp}]^{-1}[F_p - K_{ps}u_s] = 10^{-6} \begin{bmatrix} 2.0017 & -.7892 & 1.2187 & .4805 \\ -.7892 & 4.9816 & -.4805 & -.1895 \\ 1.2187 & -.4805 & 2.0017 & .7892 \\ .4805 & -.1895 & .7892 & 4.9816 \end{bmatrix}$$

$$\times \left[\begin{Bmatrix} 0 \\ -30000 \\ 0 \\ 0 \end{Bmatrix} - \begin{bmatrix} -170884 & -170884 & -172923 & 86461 \\ -170884 & -170884 & 86461 & -43231 \\ -172923 & -86461 & -170884 & 170884 \\ -86461 & -43231 & 170884 & -170884 \end{bmatrix} \begin{Bmatrix} 0 \\ -.6 \\ -.3 \\ 0 \end{Bmatrix} \right]$$

$$= \begin{Bmatrix} -.33848 \\ -.36437 \\ -.32343 \\ -.00924 \end{Bmatrix} \text{inches} \tag{2.34}$$

Equation (2.23) yields the following reactions:

$$\{F_s\} = \begin{Bmatrix} F_1 \\ F_2 \\ F_7 \\ F_8 \end{Bmatrix} = \begin{Bmatrix} 22426 \\ 20001 \\ -22425 \\ 10000 \end{Bmatrix} \text{pounds} \tag{2.35}$$

Member forces are obtained as in Example (2.3).

2.4.1 Discussion

As expected, the results obtained from the use of each technique are nearly identical (within 0.02%) with round-off error accounting for the difference.

The partitioning technique has the advantage of having to invert a 4×4 matrix rather than a 6×6. However, it requires reordering of the rows and columns of the structural stiffness matrix and a number of matrix multiplications, additions, and subtractions. The reordering is, of course, dependent on the specific displacements given. As a result, implementation of this technique in a computer program would require not only checks to determine which displacements are non-zero (as does the Chapter 1 technique), but also would require extensive bookkeeping to identify rows and columns to be interchanged and specific portions of rows and columns to be used in the other extensive mathematical operations.

In addition, for most large problems where the stiffness method would generally be used, the number of specified displacements is small in comparison to the total number of degrees of freedom. Thus, the order of the matrix to be inverted when using the partitioning technique is not much smaller than the entire reduced stiffness matrix.

Therefore, the partitioning technique, although useful from a theoretical point of view and for hand solution of small problems, is unwieldy and not advantageous to use as an algorithm for implementation in computer code.

2.5 TEMPERATURE CHANGES AND FABRICATION ERRORS

Temperature effects and fabrication errors such as an incorrect bar length are dealt with by using the concept of fixed end forces and equivalent nodal loads. Recall that we used this technique in Chapter 1 when addressing temperature changes in a one-dimensional rod element. We prevent any displacements of the nodes and determine the forces required to maintain the nodes in this undisplaced configuration (the fixed end forces). We then apply the equivalent nodal loads, which are the opposite of the fixed end forces. After determining the nodal displacements, the member forces are found by superposing the fixed end forces and the forces found using the calculated displacements.

For the truss element, as for the bar element, the fixed end forces will be directed along the axis of the member. Since the nodal forces are specified in terms of the global coordinates, we need to transform the elemental fixed end forces to the global coordinate system. This is easily done.

When we discussed vector transformations, we found that $\{\delta\} = [\beta]\{u\}$ and $\{P\} = [\beta]\{F\}$, where δ and P are expressed in terms of the elemental axes and u and F in terms of the global axes. Since $\{P\} = [\beta]\{F\}$, then $\{F\} = [\beta]^{-1}\{P\}$. Remember, however, that $[\beta]^{-1} = [\beta]^T$. Thus,

$$\{ F \} = [\beta]^T \{ P \} \tag{2.36}$$

Equation (2.36) enables us to transform the elemental equivalent nodal forces $\{P\}$ to the global equivalent forces $\{F\}$. These equivalent forces are simply the opposite of the fixed end forces resolved into the global, coordinate directions. For example, suppose that member ab in the truss shown in Figure 2-3 is fabricated 1/4 inch too short.

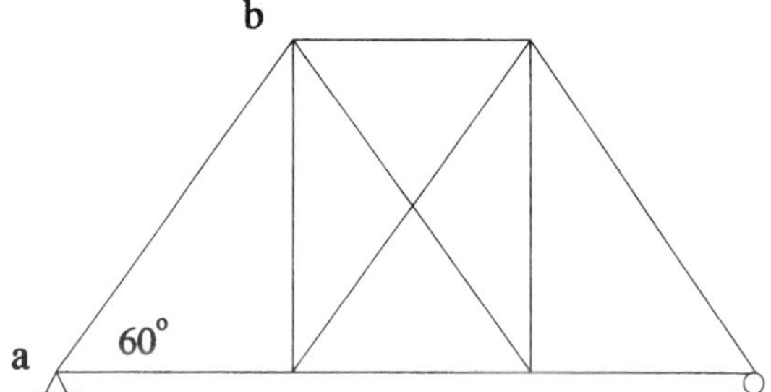

Figure 2-3

Assuming a cross-sectional area of 2 in^2, a length of 20 ft (240 in), and an $E = 29 \times 10^6$ psi, the force required to maintain zero displacements of nodes a and b is $P = EA\Delta/L = 60417\#$ (tension). The equivalent nodal forces at a and b are shown in Figure 2-4. Note that they act in a direction parallel to the axis of member ab.

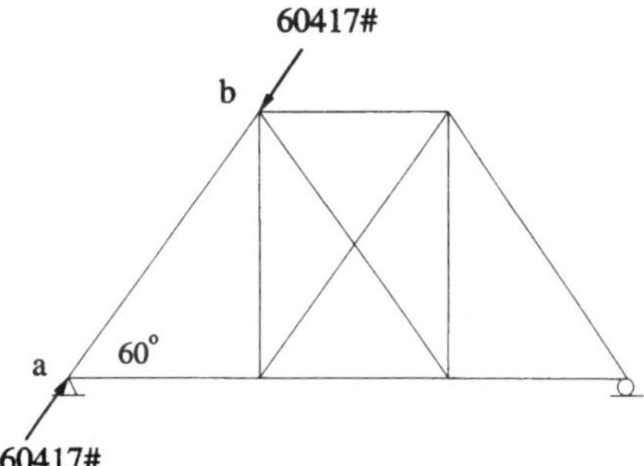

Figure 2-4

The $[\beta]$ matrix for this member is

$$[\beta]_{ab} = \begin{bmatrix} .5 & .866 & 0 & 0 \\ -.866 & .5 & 0 & 0 \\ 0 & 0 & .5 & .866 \\ 0 & 0 & -.866 & .5 \end{bmatrix} \tag{2.37}$$

Now, from equation (2.36),

$$\begin{Bmatrix} F_1 \\ F_2 \\ F_3 \\ F_4 \end{Bmatrix} = \begin{bmatrix} .5 & -.866 & 0 & 0 \\ .866 & .5 & 0 & 0 \\ 0 & 0 & .5 & -.866 \\ 0 & 0 & .866 & .5 \end{bmatrix} \begin{Bmatrix} 60417\# \\ 0 \\ -60417\# \\ 0 \end{Bmatrix} = \begin{Bmatrix} 30208.5 \\ 52322.6 \\ -30208.5 \\ -52322.6 \end{Bmatrix} \text{pounds} \tag{2.38}$$

These equivalent nodal loads are shown in Figure 2-5.

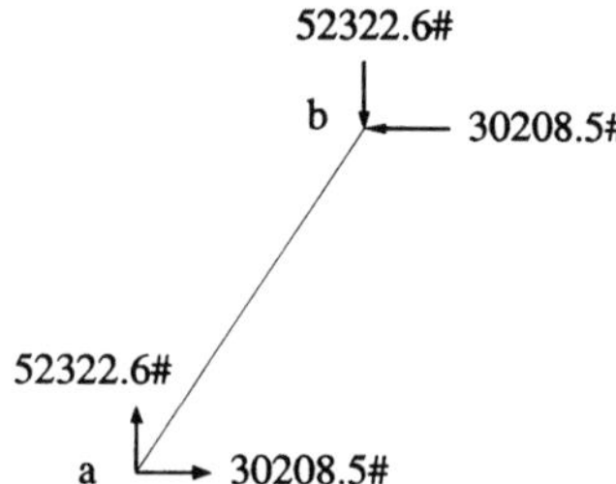

Figure 2-5

The forces shown in Figure 2-5 are now applied to the structure and nodal displacements found. After the member forces are determined, we will add the fixed end forces to member *ab*. The following examples illustrate this procedure.

Example 2.5

For the truss shown in Figure E2-3, solve for the force in member 4 if it has been fabricated 1/4 in too short. Remove the 30-kip load and the support movements.

The fixed end forces for member 4 are $EA\Delta/L = 54038\#$ (tension). The $[\beta]$ matrix for member 4 is

$$[\beta]_4 = \begin{bmatrix} .89445 & .44723 & 0 & 0 \\ -.44723 & .89445 & 0 & 0 \\ 0 & 0 & .89445 & .44723 \\ 0 & 0 & -.44723 & .89445 \end{bmatrix} \tag{2.39}$$

From equation (2.36), noting that nodes 1 and 3 are at the ends of this member, we have

$$\begin{Bmatrix} F_1 \\ F_2 \\ F_5 \\ F_6 \end{Bmatrix} = \begin{bmatrix} .89445 & -.44723 & 0 & 0 \\ .44723 & .89445 & 0 & 0 \\ 0 & 0 & .89445 & -.44723 \\ 0 & 0 & .44723 & .89445 \end{bmatrix} \begin{Bmatrix} 54038\# \\ 0 \\ -54038\# \\ 0 \end{Bmatrix} = \begin{Bmatrix} 48334 \\ 24167 \\ -48334 \\ -24167 \end{Bmatrix} \text{pounds} \tag{2.40}$$

After removing rows and columns 1, 2, 7, and 8 from the structural stiffness matrix to account for zero displacements at nodes 1 and 4, the reduced stiffness matrix becomes

$$[K]_R = \begin{bmatrix} 827140 & 84423 & -483333 & 0 \\ 84423 & 214115 & 0 & 0 \\ -483333 & 0 & 827140 & -84423 \\ 0 & 0 & -84423 & 214115 \end{bmatrix} \tag{2.41}$$

Inverting the reduced stiffness matrix and solving for the unknown nodal displacements, we find

$$\begin{Bmatrix} u_3 \\ u_4 \\ u_5 \\ u_6 \end{Bmatrix} = 10^{-6} \times \begin{bmatrix} 2.0017 & -.7892 & 1.2187 & .4805 \\ -.7892 & 4.9816 & -.4805 & -.1895 \\ 1.2187 & -.4805 & 2.0017 & .7892 \\ .4805 & -.1895 & .7892 & 4.9816 \end{bmatrix} \begin{Bmatrix} 0 \\ 0 \\ -48334 \\ -24167 \end{Bmatrix} = \begin{Bmatrix} -.07050 \\ .02780 \\ -.11582 \\ -.15854 \end{Bmatrix} \text{inches} \tag{2.42}$$

The force in member 4 is

$$\begin{Bmatrix} P_1 \\ P_2 \\ P_3 \\ P_4 \end{Bmatrix} = 216153 \begin{bmatrix} 1 & 0 & -1 & 0 \\ 0 & 0 & 0 & 0 \\ -1 & 0 & 1 & 0 \\ 0 & 0 & 0 & 0 \end{bmatrix} \begin{bmatrix} .89445 & .44723 & 0 & 0 \\ -.44723 & .89445 & 0 & 0 \\ 0 & 0 & .89445 & .44723 \\ 0 & 0 & -.44723 & .89445 \end{bmatrix} \begin{Bmatrix} 0 \\ 0 \\ -.11582 \\ -.15854 \end{Bmatrix}$$

$$+ \begin{Bmatrix} -54038 \\ 0 \\ 54038 \\ 0 \end{Bmatrix} = \begin{Bmatrix} -16319 \\ 0 \\ 16319 \\ 0 \end{Bmatrix} \text{pounds} \tag{2.43}$$

Example 2.6

Using the truss in Example 2.5, solve for the bar forces in members 2 and 4 if member 2 undergoes a temperature change of $-40°$F. Use $\alpha = 6.5 \times 10^{-6}$in/in/°F.

The fixed end forces are $P = EA\alpha(\Delta T) = 15080\#$ (tension). The equivalent nodal loads are shown in Figure E2-6. Note that a formal transformation of these forces to the global coordinate system is not necessary since the member axis is aligned with the global x-axis.

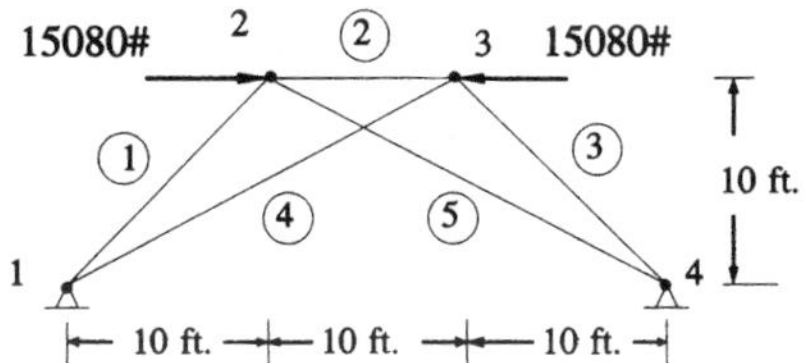

Figure E2-6

The reduced stiffness matrix was found in the previous example (equation [2.41]). Solving for the unknown displacements, we have

$$\begin{Bmatrix} u_3 \\ u_4 \\ u_5 \\ u_6 \end{Bmatrix} = 10^{-6} \times \begin{bmatrix} 2.0017 & -.7892 & 1.2187 & .4805 \\ -.7892 & 4.9816 & -.4805 & -.1895 \\ 1.2187 & -.4805 & 2.0017 & .7892 \\ .4805 & -.1895 & .7892 & 4.9816 \end{bmatrix} \begin{Bmatrix} 15080\# \\ 0 \\ -15080\# \\ 0 \end{Bmatrix} = \begin{Bmatrix} .0118 \\ -.0047 \\ -.0118 \\ -.0047 \end{Bmatrix} \text{inches} \tag{2.44}$$

The force in member 2 is

$$\begin{Bmatrix} P_1 \\ P_2 \\ P_3 \\ P_4 \end{Bmatrix} = 483333 \begin{bmatrix} 1 & 0 & -1 & 0 \\ 0 & 0 & 0 & 0 \\ -1 & 0 & 1 & 0 \\ 0 & 0 & 0 & 0 \end{bmatrix} \begin{bmatrix} 1 & 0 & 0 & 0 \\ 0 & 1 & 0 & 0 \\ 0 & 0 & 1 & 0 \\ 0 & 0 & 0 & 1 \end{bmatrix} \begin{Bmatrix} .0118 \\ -.0047 \\ -.0018 \\ -.0047 \end{Bmatrix} + \begin{Bmatrix} -15080 \\ 0 \\ 15080 \\ 0 \end{Bmatrix}$$

$$= \begin{Bmatrix} -3673 \\ 0 \\ 3673 \\ 0 \end{Bmatrix} \text{pounds} \tag{2.45}$$

The force in member 4 is

$$\begin{Bmatrix} P_1 \\ P_2 \\ P_3 \\ P_4 \end{Bmatrix} = 216153 \begin{bmatrix} 1 & 0 & -1 & 0 \\ 0 & 0 & 0 & 0 \\ -1 & 0 & 1 & 0 \\ 0 & 0 & 0 & 0 \end{bmatrix} \begin{bmatrix} .89445 & .44723 & 0 & 0 \\ -.44723 & .89445 & 0 & 0 \\ 0 & 0 & .89445 & .44723 \\ 0 & 0 & -.44723 & .89445 \end{bmatrix} \begin{Bmatrix} 0 \\ 0 \\ -.0118 \\ -.0047 \end{Bmatrix}$$

$$= \begin{Bmatrix} 2736 \\ 0 \\ -2736 \\ 0 \end{Bmatrix} \text{pounds}$$

We next discuss implementation of a computer program to solve two-dimensional truss problems.

2.6 COMPUTER FORMULATION FOR THE TRUSS

The basic outline of a truss computer program is identical to that presented in Chapter 1 for the one-dimensional bar. Since we have increased the number of degrees of freedom per node to two, and since we need to transform the elemental stiffness matrices from the elemental coordinate system to the global coordinate system, there are some additional considerations to discuss.

We first need to define the geometry of the truss. Assume that the left and right node numbers for each member have been entered in the arrays ML(I) and MR(I). After the nodal x and y coordinates X(I) and Y(I) have been entered we calculate the length of each member and the sines and cosines of the angles each member axis makes with the global x-axis.

```
L(I)=SQR((X(MR(I))-X(ML(I)))^ 2 + (Y(MR(I))-Y(ML(I)))^ 2)
S(I)=(Y(MR(I))-Y(ML(I)))/L(I)     [sine of angle]
C(I)=(X(MR(I))-X(ML(I)))/L(I)     [cosine of angle]
```

The total number of degrees of freedom is equal to twice the number of nodes, i.e., 2*NN. The displacements of each node will be identified in the following way: The x displacement of the left node of the member will be 2*ML(I)-1 and the y displacement 2*ML(I). Similarly, the x displacement of the right node will be 2*MR(I)-1 and the y displacement 2*MR(I). Notice that the examples presented in the previous section follow this convention. Since we have computed the length and sines and cosines of each member, we can write the elements of the transformed elemental stiffness matrix directly and place them in the appropriate location in the global system stiffness matrix with the following code fragment:

```
FOR I=1 TO NM  [loop on the number of members]
AK=A(I)*E(I)/L(I)
EKT(1,1)=AK*C(I)*C(I)   [element 1,1 of transformed elemental
                          stiffness from equation (2.14)]
EKT(3,3)=EKT(1,1)
EKT(2,2)=AK*S(I)*S(I)
```

```
... other stiffness elements
IJ(1)=2*ML(I)-1
IJ(2)=2*ML(I)
IJ(3)=2*MR(I)-1
IJ(4)=2*MR(I)     [identifies rows and columns in
                   the global K matrix]
FOR IR=1 TO 4
FOR IC=1 TO 4
KR=IJ(IR)  [row of global stiffness matrix]
KC=IJ(IC)  [column of global stiffness matrix]
SK(KR,KC)=SK(KR,KC)+EKT(IR,IC)  [fill global K matrix]
NEXT IC:NEXT IR
NEXT I
```

If KXRES(I) and KYRES(I) represent restraint codes (0 for no restraint, 1 for complete restraint), the order of the reduced stiffness matrix can be found using the following code fragment:

```
KSUM=0
FOR I=1 TO NN  [loop on number of nodes]
KSUM=KSUM+KXRES(I)+KYRES(I)
NEXT I
NKR=2*NN-KSUM  [order of reduced stiffness matrix]
```

As in the case of the one-dimensional rod, we now fill the KEPT array with the numbers of the rows and columns we wish to keep in the reduced stiffness matrix. In this case we have two possible restraints at each node.

```
J=0
FOR I=1 TO NN
IF(KXRES(I)>0) THEN GOTO 3020
J=J+1:KEPT(J)=2*I-1
3020 IF(KYRES(I)>0) THEN GOTO 3040
J=J+1:KEPT(J)=2*I
3040 NEXT I
```

Now we construct the reduced force and stiffness matrices.

```
FOR I=1 TO NKR
N=KEPT(I)
FR(I)=F(N)   [reduced force matrix]
FOR J=1 TO NKR
M=KEPT(J)
SKR(I,J)=SK(N,M)
NEXT J:NEXT I
```

We now invert the reduced stiffness matrix and calculate the non-zero displacements. Assuming that after inversion, SKR contains the inverse,

```
FOR I=1 TO NKR
FOR J=1 TO NKR
U(I)=U(I)+SKR(I,J)*FR(J)
NEXT J:NEXT I
```

We next place the non-zero displacements in the appropriate locations in the global displacement matrix DU(I).

```
FOR I =1 TO NKR
ND=KEPT(I)
DU(ND)=U(I)
NEXT I
```

The member forces are found by calculating P_3 for each member.

```
FOR I=1 TO NM    [loop on number of members]
K=2*ML(I)
LL=K-1
M=2*MR(I)
N=M-1
P(I)=(E(I)*A(I)/L(I))*(C(I)*(DU(N)-DU(LL))+S(I)*(DU(M)-DU(K)))
NEXT I
```

2.7 SUMMARY

In this chapter we have developed the stiffness matrix for the truss element. This necessitated using two degrees of freedom per node. In addition, since the local and global coordinate systems are not, in general, parallel, it was necessary to develop a transformation equation for the elemental stiffness matrix. This equation transformed the elemental stiffness matrix with respect to the elemental coordinate system to the global coordinate system. We could then generate the global structural stiffness matrix. We presented the topics of support movement, temperature effects, and fabrication errors. Truss examples were presented and computer formulation of the truss problem was discussed. In the next chapter we will consider two-dimensional beam and frame elements.

PROBLEMS

2.1 For the truss shown in Figure P2-1,

(1) Find the overall stiffness matrix.

(2) Using the support conditions, generate the reduced stiffness matrix.

(3) Calculate the nodal displacements and member forces.
Solve the problem by the method of joints as presented in your basic structures courses and compare with part 3 above.

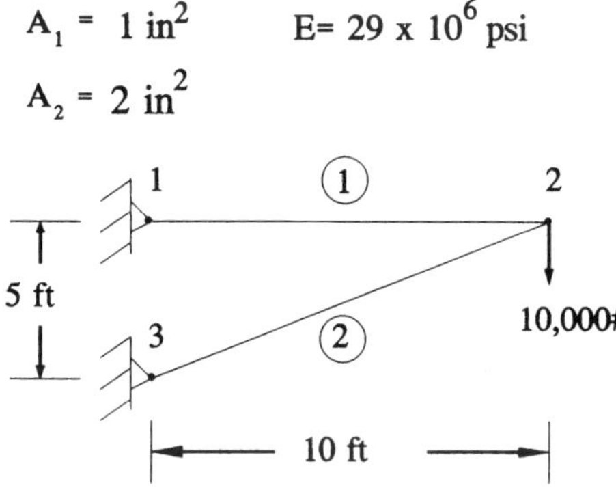

Figure P2-1

2.2 For the truss shown in Figure P2-2, find all member forces. All areas are 1 in^2 and $E = 10 \times 10^6$ psi.

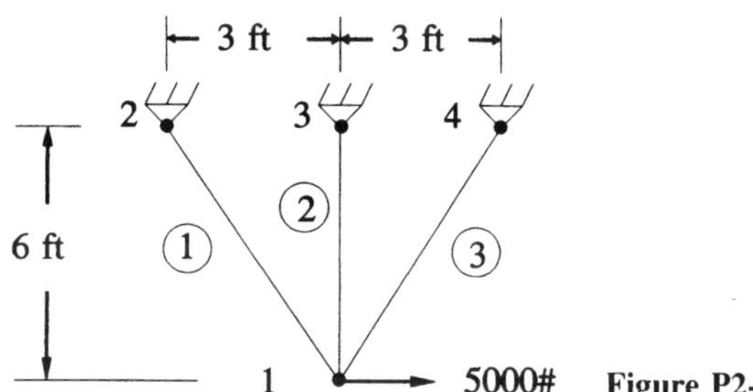

Figure P2-2

2.3 Find all member forces for the truss shown in Figure P2-3. All areas are 2 in^2 and $E = 29 \times 10^6$ psi.

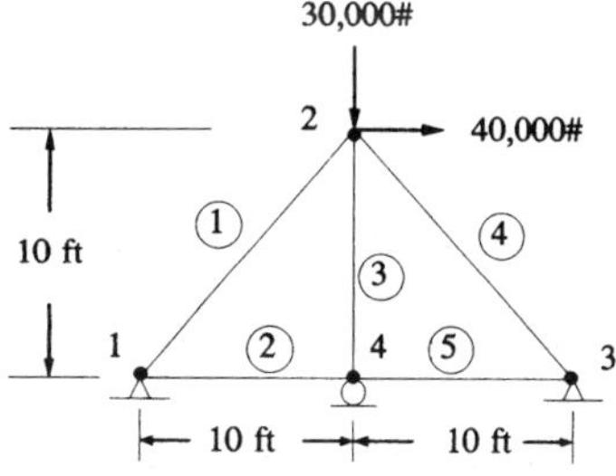

Figure P2-3

2.4 For the truss shown in Figure P2-4,

(1) Find the overall structural stiffness matrix.

(2) Generate the reduced stiffness matrix.

Calculate the nodal displacements and member forces.

The area of each member is 3 in^2.

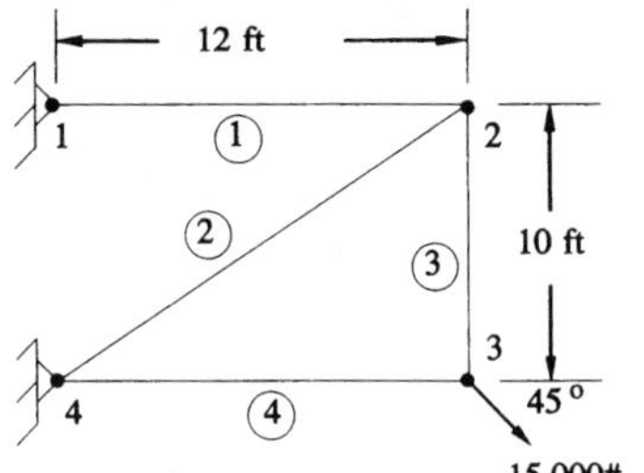

Figure P2-4

2.5 The truss in Figure 2-2 undergoes a support movement at node 4 of 3/8 in to the right in addition to the applied load of 5000#. Find the final bar forces by

(1) Using the procedure presented in Example 2.3

(2) Using the matrix partitioning technique (Section 2.4)

2.6 Member 3 of the truss in Figure P2-3 undergoes a temperature change of +40°F. Solve for the bar forces considering

(1) Only the temperature change

(2) The loads and temperature change.

Use the results of Problem 2.3 and part (1) to verify part (2) by superposition. Use $\alpha = 6.5 \times 10^{-6}{}''/''/°\text{F}$.

2.7 Find all bar forces in Figure P2-7 if member 5 is fabricated 1/4 in too short. All areas are 2 in^2 and $E = 29 \times 10^6$ psi.

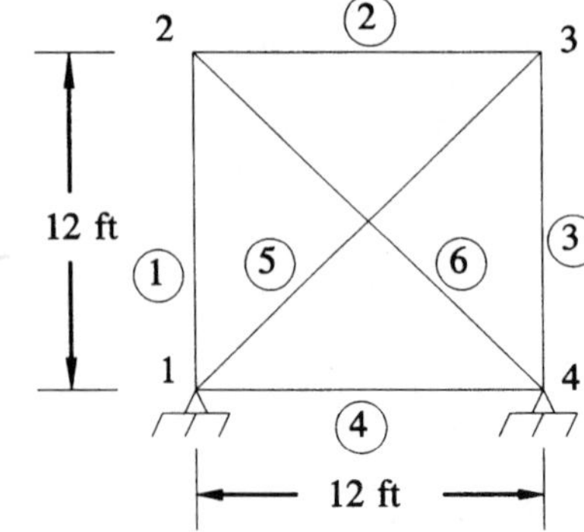

Figure P2-7

2.8 Using the computer code fragments presented in Section 2.6, develop a computer program to solve truss problems involving nodal applied loads. The program should have a capacity of 15 nodes and 20 members. Use your hand solutions to problems 2.1–2.4 to verify the operation of your program.

Use your computer program as an aid for solving the following problems:

2.9 Solve for the bar forces of the truss in Figure P2-9 if there is a vertical load of 20,000# acting down at node 4. Use $E = 29 \times 10^6$ psi. Members 4 and 5 have cross-sectional areas of 1.5 in^2. All other areas are 1 in^2.

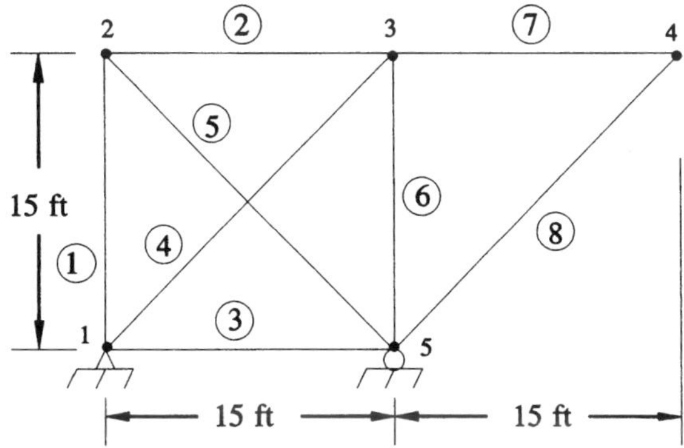

Figure P2-9

2.10 If members 2 and 7 of the truss in Figure P2-9 have a temperature increase of 40°F, find all bar forces. Use $\alpha = 6.5 \times 10^{-6}{''}/{''}/°\text{F}$.

2.11 Node 2 of the truss shown in Figure P2-11 has a horizontal load of 50,000# acting to the left and a vertical load of 25,000# acting down. Areas of members 1, 3, and 5 are 4 in^2. All other members have areas of 2 in^2. Use $E = 29 \times 10^6$ psi.

Solve for all nodal displacements and bar forces. Check equilibrium at all nodes.

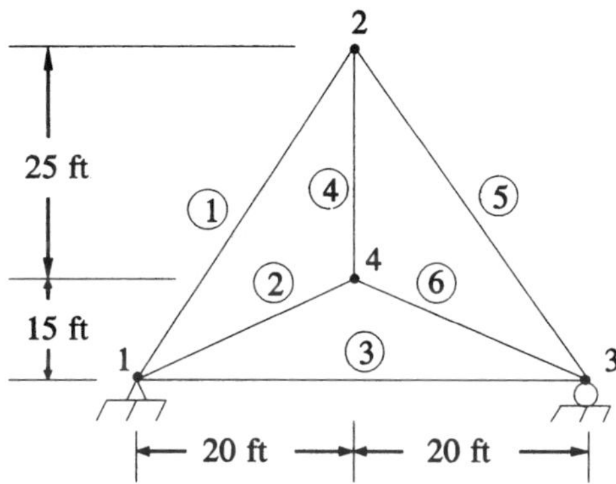

Figure P2-11

2.12 Member 4 of the truss in Figure P2-11 was fabricated 1/4 in too short. Find the bar forces for this fabrication error only.

2.13 Members 1 and 5 of the truss in Figure P2-11 are subjected to temperature changes of +30°F. Find the bar forces for these temperature changes only. Use $\alpha = 6.5 \times 10^{-6}{''}/{''}/°\text{F}$.

2.14 Bars 2, 5, and 7 of the truss shown in Figure P2-14 have areas of 5 in^2. For all other members $A = 3$ in^2. If members 4 and 6 undergo a temperature rise of 60°F, find all bar forces. $E = 29 \times 10^6$ psi and $\alpha = 6.5 \times 10^{-6}{''}/{''}/°\text{F}$. Check equilibrium of all nodes.

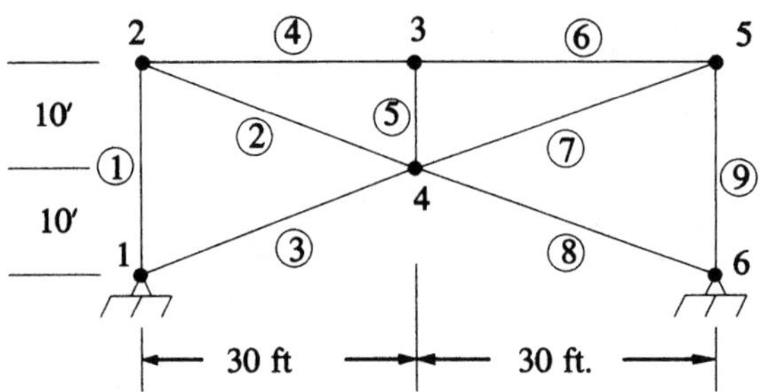

Figure P2-14

2.15 Horizontal loads of 40,000# are acting to the right at nodes 2 and 5 of the truss in Figure P2-14. Solve for all bar forces.

2.16 Solve problem P2.15 where, in addition to the applied loads, member 5 is fabricated 1/8 in too long.

CHAPTER 3

ANALYSIS OF TWO-DIMENSIONAL BEAMS AND FRAMES

3.1 INTRODUCTION

Beam and frame elements play major roles in structural analysis and design. They provide flexural rigidity, a property that the two-force truss element does not have. In fact, after analyzing a truss for axial forces due to nodal loads, which are often determined using the concept of tributary area, we then must separately consider the non-nodal loads that cause bending in the truss member. A roof truss is an example of a structure where this procedure is required. For example, consider the simple roof truss shown in Figure 3-1.

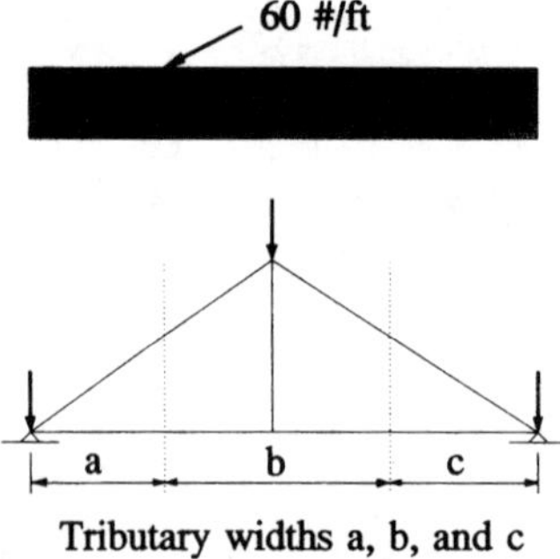

Figure 3-1 Roof truss showing tributary widths.

Assume that the total load for which the truss must be designed is 60 lb per foot of span. The nodal loads that are acting on the truss would be determined by using the loads contained in the indicated tributary areas. An analysis of the truss would next be performed and the member axial forces found. However, the top chord members of this truss are actually subjected to a distributed load that causes these members to act as beams in bending. An analysis that treats these members as beams is next performed and the results superposed on the nodal force solution in order to find the total axial forces and bending moments for which each member must be designed.

Figure 3-2 shows sketches of both a beam and a frame element.

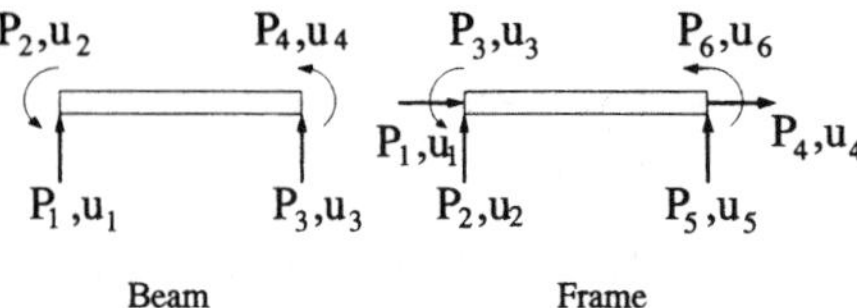

Figure 3-2 Beam and frame elements.

Note that the beam element has two degrees of freedom at each end: a rotation about an axis perpendicular to the plane of the beam and a translation perpendicular to the axis of the beam. Axial deformation is neglected. In contrast to the beam element, the frame element includes axial deformation at each end in addition to the beam deformations. It therefore has three degrees of freedom at each end, or node, of the element. One additional important difference deals with the orientation of the elements when forming a structure. The beam elements have their longitudinal axis aligned as did the one-dimensional bar element. However, the axis of the two-dimensional frame member can have any orientation in the plane of the structure.

Since the frame element has axial deformation added to the deformations of the beam element, and since there is no coupling between the axial and flexural deformations (for small displacements), we will first derive the stiffness matrix for the beam element and then add the effects of axial deformation to obtain the elemental stiffness matrix for the frame element.

Note, also, that there is no coordinate transformation required for the beam element, since, as in the case of the one-dimensional bar, all elemental axes are aligned. However, since the frame element can have any orientation in its plane, the development of the $[\beta]$ transformation matrix will be necessary.

3.2 THE BEAM ELEMENTAL STIFFNESS MATRIX

Since the beam element has a total of four degrees of freedom we can write

$$\begin{Bmatrix} P_1 \\ P_2 \\ P_3 \\ P_4 \end{Bmatrix} = \begin{bmatrix} k_{11} & k_{12} & k_{13} & k_{14} \\ k_{21} & k_{22} & k_{23} & k_{24} \\ k_{31} & k_{32} & k_{33} & k_{34} \\ k_{41} & k_{42} & k_{43} & k_{44} \end{bmatrix} \begin{Bmatrix} u_1 \\ u_2 \\ u_3 \\ u_4 \end{Bmatrix} \tag{3.1}$$

In equation (3.1), the $\{P\}$ matrix represents the elemental forces at each end of the member, the k_{ij}'s the elements of the beam stiffness matrix, and the $\{u\}$ matrix the displacements of the nodes of the member. Note that since the elemental displacements are in the same directions as the global displacements, we have used $\{u\}$ rather than $\{\delta\}$ for the elemental nodal displacements. We will not be able to do this for the frame element.

Using the definition of stiffness, in order to determine the elements of the stiffness matrix, we must introduce a unit displacement at one and only one degree of freedom at a time and determine the forces corresponding to this displacement pattern. We will use the slope-deflection equations to accomplish this task.

Consider the beam in Figure 3-3:

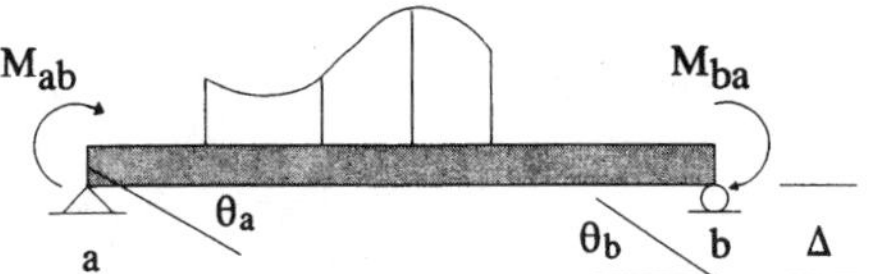

Figure 3-3 Positive forces and displacements for the beam element.

In general, the slope-deflection equations for the beam shown in Figure 3-3 can be written (see Appendix C):

$$M_{ab} = 2EI/L\ [2\theta_a + \theta_b - 3\Delta/L] - 2EI/L\,[2\theta_a^L + \theta_b^L] \tag{3.2}$$

$$M_{ba} = 2EI/L\ [2\theta_b + \theta_a - 3\Delta/L] - 2EI/L\,[2\theta_b^L + \theta_a^L] \tag{3.3}$$

In the above equations, M_{ab} and M_{ba} are the moments at the a and b ends of the beam, taken positive clockwise; θ_a and θ_b are the total rotation angles at each end, also taken positive when clockwise; Δ represents the relative displacement of one end of the beam with respect to the other perpendicular to the axis of the beam, taken positive when the chord of the member rotates clockwise; and θ_a^L and θ_b^L are the angles of rotation at the ends of a simply supported beam due to applied lateral loads, also positive clockwise. Note that the last terms in these equations represent expressions for the fixed-end moments due to loads. Note that we can calculate these fixed-end moments by evaluating rotations at the ends of a simply supported beam due to the applied loads. We now proceed with the derivation of the elements of the beam stiffness matrix.

Figure 3-4 shows a beam where the displacement u_1 is equal to unity.

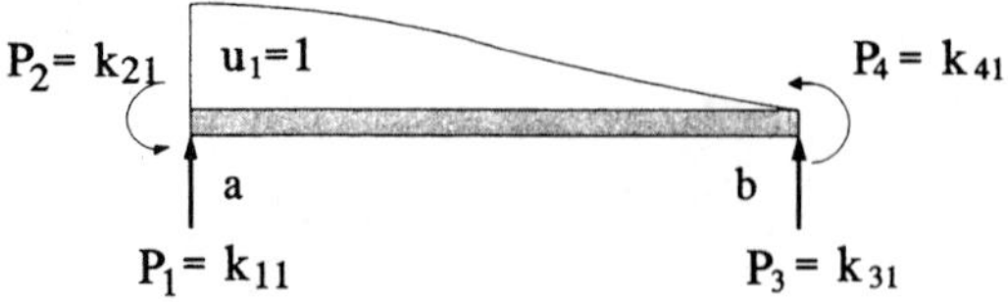

Figure 3-4 Beam deformation for $u_1 = 1$.

Writing the slope-deflection equations for this case yields

$$M_{ab} = 2EI/L\ [0 + 0 - 3(1)/L] = -6EI/L^2 \tag{3.4}$$

$$M_{ba} = 2EI/L\,[0 + 0 - 3(1)/L] = -6EI/L^2 \tag{3.5}$$

Note that $\Delta = u_1$ is positive, giving negative end moments, which indicates that these moments act in a counterclockwise direction. This direction for moments is positive with respect to the elemental coordinate system shown in Figure 3-2. Thus, $k_{21} = 6EI/L^2$ and $k_{41} = 6EI/L^2$. Summing moments about the right end of the beam yields $k_{11} = 12EI/L^3$. Vertical equilibrium gives $k_{31} = -12EI/L^3$.

We next introduce the displacement $u_2 = 1$ as shown in Figure 3-5.

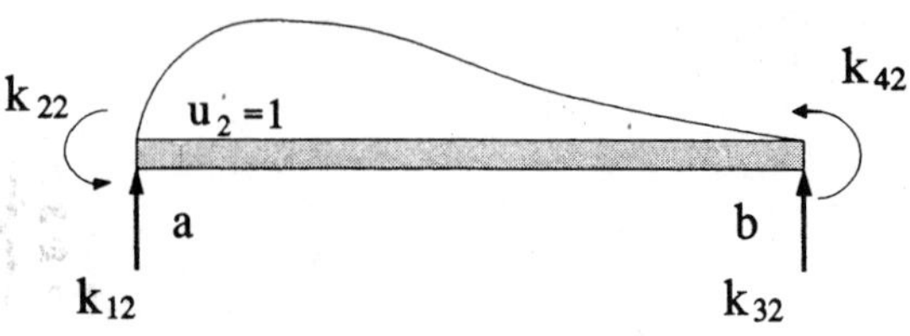

Figure 3-5 Beam deflection for $u_2 = 1$.

The slope-deflection equations yield

$$M_{ab} = 2EI/L\,[\,2(-1) + 0 - 0\,] = -4EI/L \tag{3.6}$$

$$M_{ba} = 2EI/L\,[\,0 + (-1) - 0\,] = -2EI/L \tag{3.7}$$

These equations give $k_{22} = 4EI/L$ and $k_{42} = 2EI/L$. Summing moments about one end and writing the vertical equilibrium equations result in $k_{12} = 6EI/L^2$ and $k_{32} = -6EI/L^2$.

The remaining two columns of the stiffness matrix are found by individually introducing unit displacements at the right end of the member. The final form for equation (3.1) becomes

$$\begin{Bmatrix} P_1 \\ P_2 \\ P_3 \\ P_4 \end{Bmatrix} = \begin{bmatrix} 12EI/L^3 & 6EI/L^2 & -12EI/L^3 & 6EI/L^2 \\ 6EI/L^2 & 4EI/L & -6EI/L^2 & 2EI/L \\ -12EI/L^3 & -6EI/L^2 & 12EI/L^3 & -6EI/L^2 \\ 6EI/L^2 & 2EI/L & -6EI/L^2 & 4EI/L \end{bmatrix} \begin{Bmatrix} u_1 \\ u_2 \\ u_3 \\ u_4 \end{Bmatrix} \tag{3.8}$$

You should verify the last two columns of this matrix by writing the slope-deflection and equilibrium equations for unit displacements specified at the right end of the beam.

3.3 STIFFNESS MATRIX FOR THE TWO-DIMENSIONAL FRAME ELEMENT

Consider the frame element shown in Figure 3-6.

Figure 3-6 Positive displacements for the frame elements.

Notice that we have labeled the displacements δ_i in order to distinguish the elemental coordinates from the global coordinates. This is necessary since the frame element can be oriented in any direction in the plane of the structure. The axial forces and deformations that we must add are P_1, δ_1, and P_4, δ_4. Thus, the axial stiffness terms must be included by adding row and column numbers 1 and 4. Furthermore, since the axial terms are uncoupled from the flexural terms—that is, moments and shears are not affected by the axial forces and vice versa—the only non-zero elements in these rows and columns will be those that multiply the displacements δ_1 and δ_4.

Using the results for the one-dimensional bar element (for the axial effects), we can expand the beam stiffness matrix to obtain

$$\begin{Bmatrix} P_1 \\ P_2 \\ P_3 \\ P_4 \\ P_5 \\ P_6 \end{Bmatrix} = \begin{bmatrix} EA/L & 0 & 0 & -EA/L & 0 & 0 \\ 0 & 12EI/L^3 & 6EI/L^2 & 0 & -12EI/L^3 & 6EI/L^2 \\ 0 & 6EI/L^2 & 4EI/L & 0 & -6EI/L^2 & 2EI/L \\ -EA/L & 0 & 0 & EA/L & 0 & 0 \\ 0 & -12EI/L^3 & -6EI/L^2 & 0 & 12EI/L^3 & -6EI/L^2 \\ 0 & 6EI/L^2 & 2EI/L & 0 & -6EI/L^2 & 4EI/L \end{bmatrix} \begin{Bmatrix} \delta_1 \\ \delta_2 \\ \delta_3 \\ \delta_4 \\ \delta_5 \\ \delta_6 \end{Bmatrix} \tag{3.9}$$

Equation (3.9) is the elemental stiffness matrix for the frame element in terms of the local or elemental coordinate system. As in the case of the truss element, we must develop the $[\beta]$ transformation matrix for the frame element.

3.4 THE TRANSFORMATION MATRIX FOR THE FRAME ELEMENT

Consider the frame element shown in Figure 3-7.

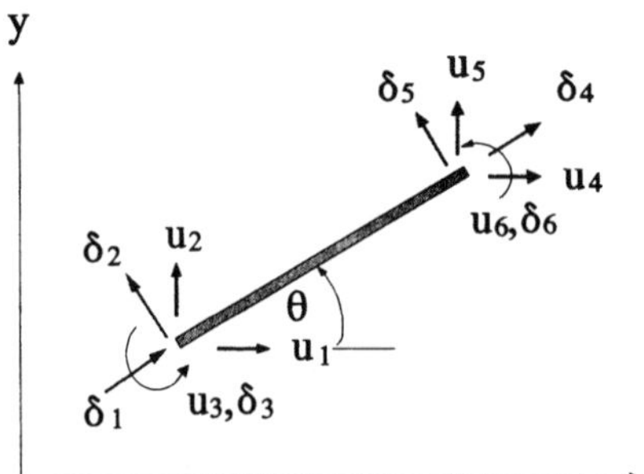

Figure 3-7 Local and global displacements for the frame element.

The relationships between the elemental translational displacements δ_1, δ_2, δ_4, and δ_5 and the global displacements u_1, u_2, u_4, and u_5 are identical to those for the truss element. Also, the rotational displacements δ_3 and δ_6 are the same as u_3 and u_6 since we are dealing with a two-dimensional element. That is, the local z axis has a direction out of the plane of the element as does the global z axis. We can therefore write

$$\begin{aligned}
\delta_1 &= u_1 \cos\theta + u_2 \sin\theta \\
\delta_2 &= u_2 \cos\theta - u_1 \sin\theta \\
\delta_3 &= u_3 \\
\delta_4 &= u_4 \cos\theta + u_5 \sin\theta \\
\delta_5 &= u_5 \cos\theta - u_4 \sin\theta \\
\delta_6 &= u_6
\end{aligned}$$

Since $\{\delta\} = [\beta]\{u\}$, rewriting the equations above in matrix form yields the following beta matrix:

$$[\beta] = \begin{bmatrix} \cos\theta & \sin\theta & 0 & 0 & 0 & 0 \\ -\sin\theta & \cos\theta & 0 & 0 & 0 & 0 \\ 0 & 0 & 1 & 0 & 0 & 0 \\ 0 & 0 & 0 & \cos\theta & \sin\theta & 0 \\ 0 & 0 & 0 & -\sin\theta & \cos\theta & 0 \\ 0 & 0 & 0 & 0 & 0 & 1 \end{bmatrix} \tag{3.10}$$

As in the case of the truss element,

$$[k_e]_{global} = [\beta]^T [k_e]_{element} [\beta] \tag{3.11}$$

Using equations (3.9) and (3.10) in equation (3.11) and expanding, we find the following elemental stiffness matrix for the frame element transformed to the global coordinate system:

$$
[k_e]_{system} = \begin{bmatrix}
\begin{matrix}(EA/L)C^2\\+(12EI/L^3)S^2\end{matrix} & \begin{matrix}(EA/L)SC\\-(12EI/L^3)SC\end{matrix} & -(6EI/L^2)S & \begin{matrix}-(EA/L)C^2\\-(12EI/L^3)S^2\end{matrix} & \begin{matrix}-(EA/L)SC\\+(12EI/L^3)SC\end{matrix} & -(6EI/L^2)S \\
- & \begin{matrix}(EA/L)S^2\\+(12EI/L^3)C^2\end{matrix} & (6EI/L^2)C & \begin{matrix}-(EA/L)SC\\+(12EI/L^3)CS\end{matrix} & \begin{matrix}-(EA/L)S^2\\-(12EI/L^3)C^2\end{matrix} & (6EI/L^2)C \\
- & - & 4EI/L & (6EI/L^2)S & -(6EI/L^2)C & 2EI/L \\
- & symmetric & - & \begin{matrix}(EA/L)C^2\\+(12EI/L^3)S^2\end{matrix} & \begin{matrix}(EA/L)SC\\-(12EI/L^3)SC\end{matrix} & (6EI/L^2)S \\
- & - & - & - & \begin{matrix}(EA/L)S^2\\+12(EI/L^3)C^2\end{matrix} & -(6EI/L^2)C \\
- & - & - & - & - & 4EI/L
\end{bmatrix}
\tag{3.12}
$$

In equation (3.12), the S and C represent the sine and cosine of the angle θ between the global x axis and local x axis, positive counterclockwise.

As before, after each elemental stiffness has been transformed to the global coordinate system, we can combine them to form the global structural stiffness matrix. The general procedure for solving the equations and calculating member forces is the same as described previously. We next consider simple beam and frame examples.

3.5 EXAMPLE BEAM AND FRAME PROBLEMS

Example 3.1

Consider the beam shown in Figure E3-1a.

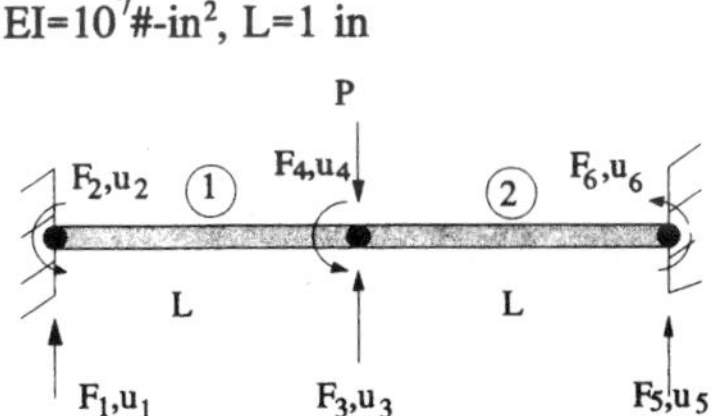

Figure E3-1a Two-element beam structure.

Remember that for the beam element, the elemental and global forces and displacements are in the same direction. Using the values for EI and L given, using equation (3.8) we can write for element 1,

$$\begin{Bmatrix} F_1 \\ F_2 \\ F_3 \\ F_4 \end{Bmatrix} = 10^7 \begin{bmatrix} 12 & 6 & -12 & 6 \\ 6 & 4 & -6 & 2 \\ -12 & -6 & 12 & -6 \\ 6 & 2 & -6 & 4 \end{bmatrix} \begin{Bmatrix} u_1 \\ u_2 \\ u_3 \\ u_4 \end{Bmatrix} \tag{3.13}$$

For element number 2 we have

$$\begin{Bmatrix} F_3 \\ F_4 \\ F_5 \\ F_6 \end{Bmatrix} = 10^7 \begin{bmatrix} 12 & 6 & -12 & 6 \\ 6 & 4 & -6 & 2 \\ -12 & -6 & 12 & -6 \\ 6 & 2 & -6 & 4 \end{bmatrix} \begin{Bmatrix} u_3 \\ u_4 \\ u_5 \\ u_6 \end{Bmatrix} \tag{3.14}$$

In the case of the beam element, the $\{P\}$ and $\{F\}$ matrices as well as the $\{\delta\}$ and $\{u\}$ matrices are interchangeable. It is important, however, to write the individual force-displacement relationships keeping in mind that the first two forces and displacements are at the left end of the member and the second two are at the right end of the member. Combining to form the structural stiffness matrix we have

$$\begin{Bmatrix} F_1 \\ F_2 \\ F_3 \\ F_4 \\ F_5 \\ F_6 \end{Bmatrix} = 10^7 \begin{bmatrix} 12 & 6 & -12 & 6 & 0 & 0 \\ 6 & 4 & -6 & 2 & 0 & 0 \\ -12 & -6 & 24 & 0 & -12 & 6 \\ 6 & 2 & 0 & 8 & -6 & 2 \\ 0 & 0 & -12 & -6 & 12 & -6 \\ 0 & 0 & 6 & 2 & -6 & 4 \end{bmatrix} \begin{Bmatrix} u_1 \\ u_2 \\ u_3 \\ u_4 \\ u_5 \\ u_6 \end{Bmatrix} \tag{3.15}$$

Noting that $u_1 = u_2 = u_5 = u_6 = 0$, we eliminate the rows and columns associated with these displacements. The reduced set of equations becomes

$$\begin{Bmatrix} F_3 \\ F_4 \end{Bmatrix} = 10^7 \begin{bmatrix} 24 & 0 \\ 0 & 8 \end{bmatrix} \begin{Bmatrix} u_3 \\ u_4 \end{Bmatrix} \tag{3.16}$$

Solving for the displacements u_3 and u_4 yields

$$u_3 = -P/24 \times 10^7 \text{ and } u_4 = 0.$$

The member forces are calculated next using the elemental force-displacement relationships.

Member 1:

$$\begin{Bmatrix} P_1 \\ P_2 \\ P_3 \\ P_4 \end{Bmatrix} = 10^7 \begin{bmatrix} 12 & 6 & -12 & 6 \\ 6 & 4 & -6 & 2 \\ -12 & -6 & 12 & -6 \\ 6 & 2 & -6 & 4 \end{bmatrix} \begin{Bmatrix} 0 \\ 0 \\ -P/(24 \times 10^7) \\ 0 \end{Bmatrix} = \begin{Bmatrix} P/2 \\ P/4 \\ -P/2 \\ P/4 \end{Bmatrix} \tag{3.17}$$

Member 2:

$$\begin{Bmatrix} P_1 \\ P_2 \\ P_3 \\ P_4 \end{Bmatrix} = 10^7 \begin{bmatrix} 12 & 6 & -12 & 6 \\ 6 & 4 & -6 & 2 \\ -12 & -6 & 12 & -6 \\ 6 & 2 & -6 & 4 \end{bmatrix} \begin{Bmatrix} -P/(24 \times 10^7) \\ 0 \\ 0 \\ 0 \end{Bmatrix} = \begin{Bmatrix} -P/2 \\ -P/4 \\ P/2 \\ -P/4 \end{Bmatrix} \tag{3.18}$$

Note that in the above equations we have used the $\{P\}$ matrix in order to emphasize that these are *elemental* forces. Free-body diagrams of the members and node 2 are shown in Figure E3-1b. The actual directions of the forces are shown. That is, positive forces act in positive elemental coordinate directions and negative forces act in the negative coordinate directions.

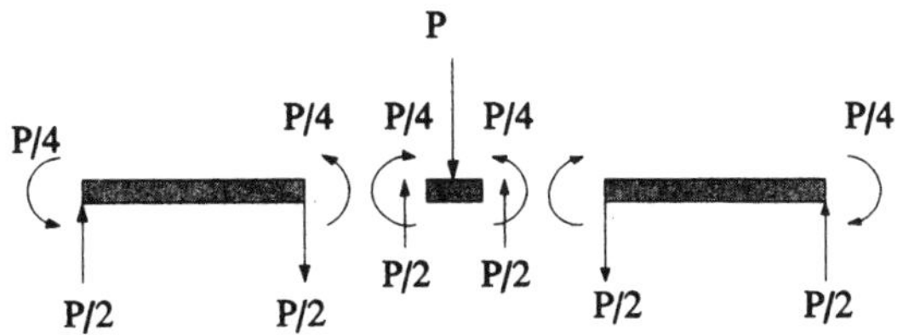

Figure E3-1b Member and nodal forces.

The bending moments at the fixed ends of the beams check with the value of $PL/8$ for the case of a concentrated force applied at the center of the beam (note that $L = 2L$ in this example).

Since the sign convention we are using for the matrix formulation is different from the common strength-of-materials sign convention, it is important to sketch a figure such as Figure E3-1b before attempting to construct shearing-force and bending-moment diagrams. It also allows an equilibrium check to be performed on the structure.

Using the forces shown in Figure E3-1b, we can construct the diagrams shown below ($L = 1$).

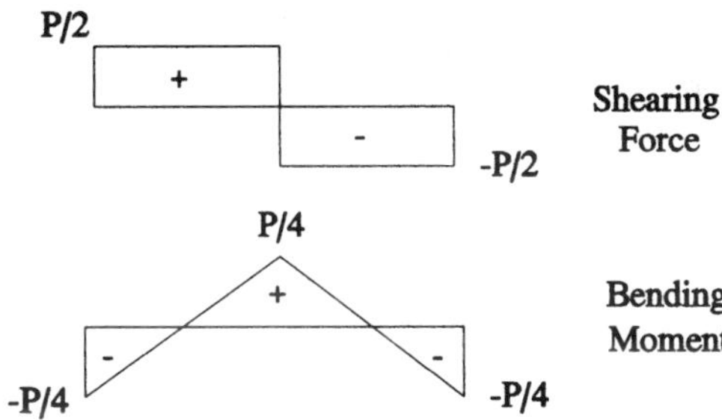

We next consider a simple frame.

Example 3.2

Consider the frame shown in Figure E3-2a.

Keeping all units in terms of pounds and inches we find:

$$(EA/L)_1 = 4.166 \times 10^6, (EI/L^2)_1 = 28935.19, (EI/L^3)_1 = 1004.69$$

$$(EA/L)_2 = 7.14286 \times 10^6, (EI/L^2)_2 = 425170, (EI/L^3)_2 = 5061.55$$

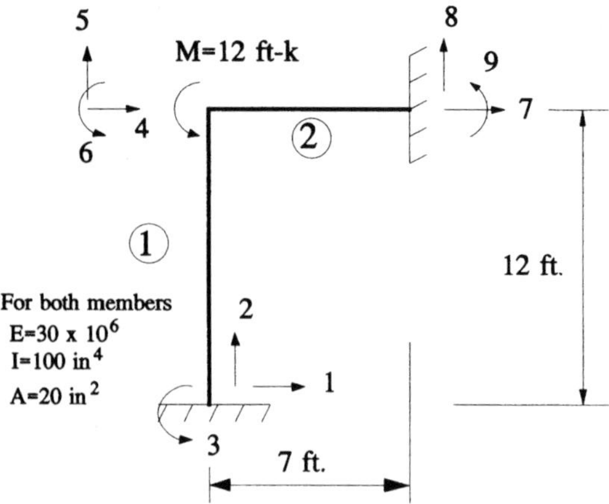

Figure E3-2a Example frame problem.

Since only displacements 4, 5, and 6 are non-zero, we need only to calculate those terms in each transformed elemental stiffness matrix that multiply these displacements. If we consider the left end of member 1 to be the fixed end, we need to calculate the lower 3×3 portion of the 6×6 transformed stiffness matrix for this member. Similarly, only the upper left 3×3 of the stiffness matrix for member 2 needs to be calculated assuming that its right end is fixed.

For member 1, $\theta = 90^o$, $S = 1, C = 0$. We find using equation (3.12),

$$\begin{Bmatrix} F_4 \\ F_5 \\ F_6 \end{Bmatrix} = \begin{bmatrix} 12{,}056.3 & 0 & 868{,}055 \\ 0 & 4.166 \times 10^6 & 0 \\ 868{,}055 & 0 & 83.33 \times 10^6 \end{bmatrix} \begin{Bmatrix} u_4 \\ u_5 \\ u_6 \end{Bmatrix} \tag{3.19}$$

For member 2, $\theta = 0^o$, $S = 0$, $C = 1$. From equation (3.12),

$$\begin{Bmatrix} F_4 \\ F_5 \\ F_6 \end{Bmatrix} = \begin{bmatrix} 7.1429 \times 10^6 & 0 & 0 \\ 0 & 60{,}738.6 & 2.551 \times 10^6 \\ 0 & 2.551 \times 10^6 & 142.86 \times 10^6 \end{bmatrix} \begin{Bmatrix} u_4 \\ u_5 \\ u_6 \end{Bmatrix} \tag{3.20}$$

Combining, we find

$$\begin{Bmatrix} F_4 \\ F_5 \\ F_6 \end{Bmatrix} = \begin{Bmatrix} 0 \\ 0 \\ 144{,}000 \end{Bmatrix} = \begin{bmatrix} 7.155 \times 10^6 & 0 & 868{,}055 \\ 0 & 4.227 \times 10^6 & 2.551 \times 10^6 \\ 868{,}055 & 2.551 \times 10^6 & 226.19 \times 10^6 \end{bmatrix} \begin{Bmatrix} u_4 \\ u_5 \\ u_6 \end{Bmatrix} \tag{3.21}$$

Solving for the displacements we have

$$u_4 = -.000078 \text{ in}, u_5 = -.000387 \text{ in, and } u_6 = .000641 \text{ rads.}$$

Now, for member forces, $\{\delta\} = [\beta]\{u\}$, $\{P\} = [k_e]\{\delta\}$. Therefore, $\{P\} = [k_e][\beta]\{u\}$, where $[k_e]$ and $[\beta]$ are given by equations (3.9) and (3.10), respectively.

For member 1,

$$\begin{Bmatrix} \delta_1 \\ \delta_2 \\ \delta_3 \\ \delta_4 \\ \delta_5 \\ \delta_6 \end{Bmatrix} = \begin{bmatrix} 0 & 1 & 0 & 0 & 0 & 0 \\ -1 & 0 & 0 & 0 & 0 & 0 \\ 0 & 0 & 1 & 0 & 0 & 0 \\ 0 & 0 & 0 & 0 & 1 & 0 \\ 0 & 0 & 0 & -1 & 0 & 0 \\ 0 & 0 & 0 & 0 & 0 & 1 \end{bmatrix} \begin{Bmatrix} 0 \\ 0 \\ 0 \\ -.000078 \\ -.000387 \\ .000641 \end{Bmatrix} = \begin{Bmatrix} 0 \\ 0 \\ 0 \\ -.000387 \\ .000078 \\ .000641 \end{Bmatrix} \tag{3.22}$$

$$\begin{Bmatrix} P_1 \\ P_2 \\ P_3 \\ P_4 \\ P_5 \\ P_6 \end{Bmatrix} = \begin{bmatrix} - & - & - & -4.166 \times 10^6 & 0 & 0 \\ - & - & - & 0 & -12{,}053.3 & 868{,}055 \\ - & - & - & 0 & -868{,}055 & 41.67 \times 10^6 \\ - & - & - & 4.166 \times 10^6 & 0 & 0 \\ - & - & - & 0 & 12{,}056.3 & -868{,}055 \\ - & - & - & 0 & -868{,}055 & 83.33 \times 10^6 \end{bmatrix} \begin{Bmatrix} 0 \\ 0 \\ 0 \\ -.000387 \\ .000078 \\ .000641 \end{Bmatrix} = \begin{Bmatrix} 1612\# \\ 555.5\# \\ 26643\ in.\text{-}\# \\ -1612\# \\ -555.5 \\ 53347 in.\text{-}\# \end{Bmatrix} \tag{3.23}$$

For member 2, the $[\beta]$ matrix is an identity matrix since the member is horizontal. As in the case of member 1, including only the terms needed, we find

$$\begin{Bmatrix} P_1 \\ P_2 \\ P_3 \\ P_4 \\ P_5 \\ P_6 \end{Bmatrix} = \begin{bmatrix} 7.1428 \times 10^6 & 0 & 0 & - & - & - \\ 0 & 60{,}738.6 & 2.551 \times 10^6 & - & - & - \\ 0 & 2.551 \times 10^6 & 142.86 \times 10^6 & - & - & - \\ -7.1428 \times 10^6 & 0 & 0 & - & - & - \\ 0 & -60{,}738.6 & -2.551 \times 10^6 & - & - & - \\ 0 & 2.551 \times 10^6 & 71.428 \times 10^6 & - & - & - \end{bmatrix} \begin{Bmatrix} -.000078 \\ -.000387 \\ .000642 \\ 0 \\ 0 \\ 0 \end{Bmatrix} = \begin{Bmatrix} -557\ \# \\ 1612\ \# \\ 91374\ in.\text{-}\# \\ 557\ \# \\ -1612\ \# \\ 44798\ in.\text{-}\# \end{Bmatrix} \tag{3.24}$$

Figure E3-2b shows the member forces and joint moments after converting the bending moments to units of ft-k. Allowing for slight round-off error, we see that equilibrium of the frame is satisfied.

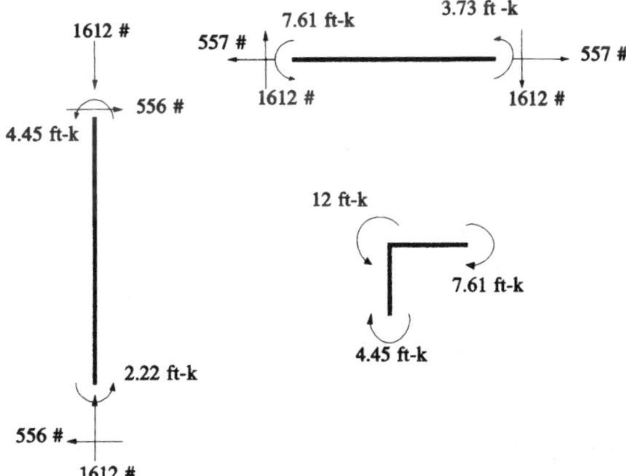

Figure E3-2b Member forces.

3.6 NON-NODAL LOADS

Consider the two-element simply supported beam shown in Figure 3-8.

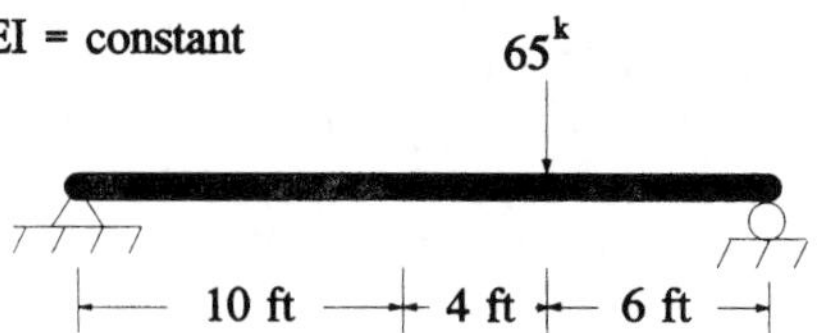

Figure 3-8 Two-element simply supported beam.

Suppose that the joints are locked before the loads are applied. This is what we do when using the moment distribution method for analyzing beams and frames. The fixed end forces acting *on the members* and a sketch of the deflection diagram are shown in Figure 3-9. Note that the fixed end moments are given by Pab^2/L^2 and Pa^2b/L^2 for a concentrated load positioned as shown where a and b are the distances to the load from the left and right ends of the member, respectively.

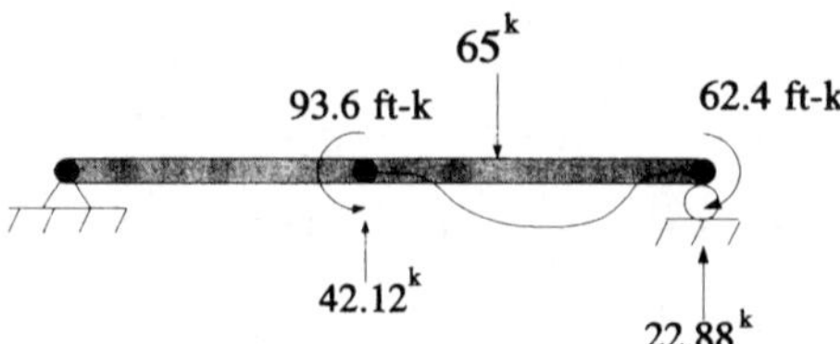

Figure 3-9 Fixed end forces.

Reversing the forces shown in Figure 3-9 gives us the equivalent nodal loads acting on the structure. These are shown in Figure 3-10. Note that the joints are again locked before the application of the loads.

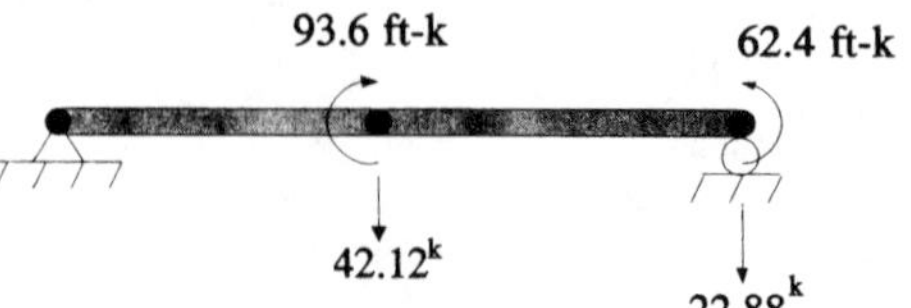

Figure 3-10 Equivalent nodal loads.

If the nodes in Figure 3-10 are now unlocked and the displacements found, they will be the correct displacements of the original structure since the loads are those applied to the nodes by the members—the equivalent nodal loads. This must be true since the superposition of the forces and nodal displacements shown in Figures 3-9 and 3-10 clearly will represent the original structure of Figure 3-8.

However, to determine the *member forces* we must add those shown in Figure 3-9 to those determined from the solution to the problem shown in Figure 3-10. This procedure is identical to that described for the one-dimensional bar element.

Of course, for many relatively simple loading conditions, the fixed end forces have been tabulated. Shown below are the fixed end moments and reactions for a few of the most common loading conditions.

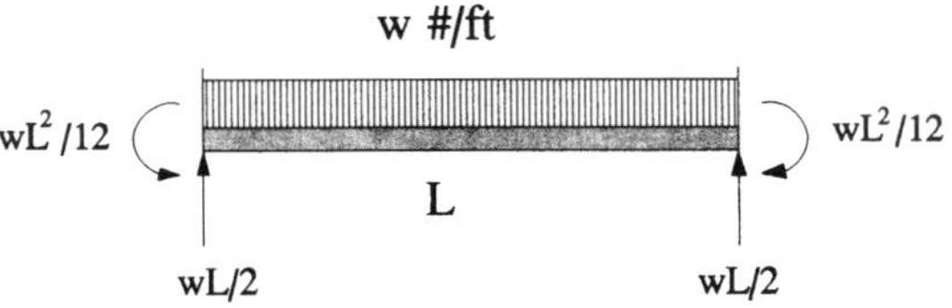

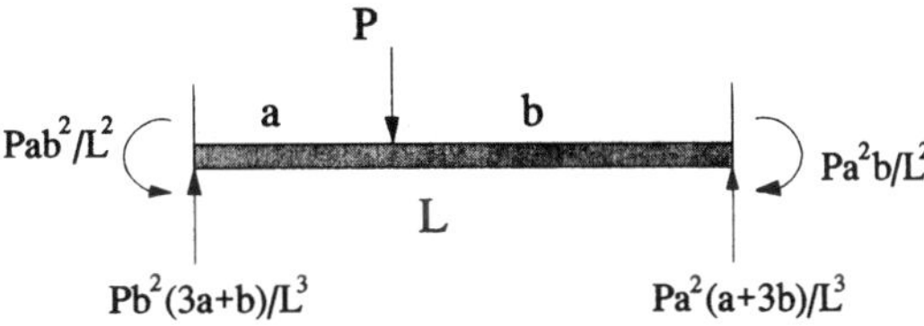

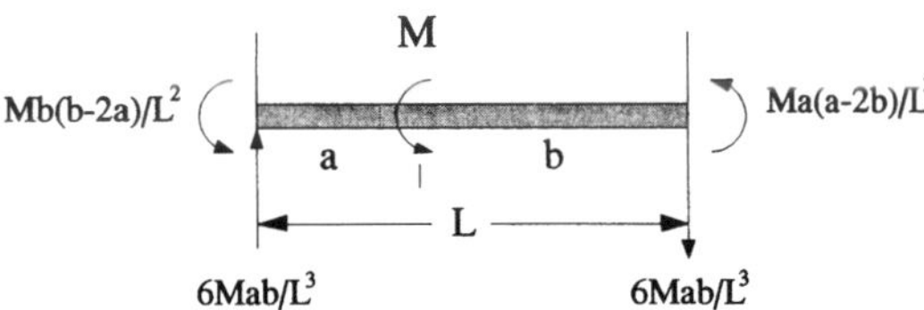

Example 3.3

Consider the beam in Figure 3-11.

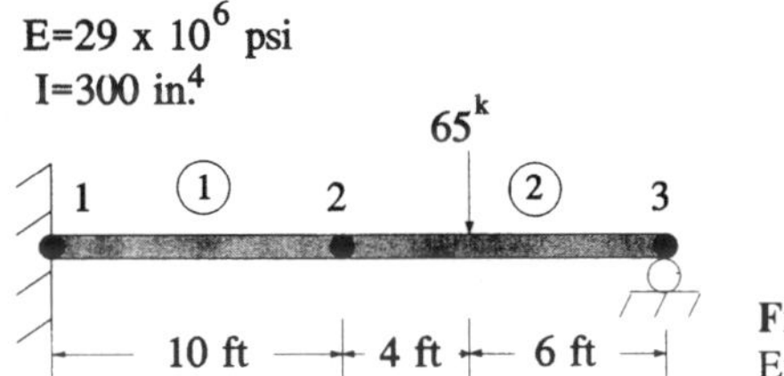

Figure 3-11 Structure for Example 3.3.

Using equation (3.8) for both members, we find:

Member 1:

$$\begin{Bmatrix} P_1 \\ P_2 \\ P_3 \\ P_4 \end{Bmatrix} = \begin{bmatrix} 60{,}416.66 & 3.625 \times 10^6 & -60{,}416.66 & 3.625 \times 10^6 \\ 3.625 \times 10^6 & 290 \times 10^6 & -3.625 \times 10^6 & 145 \times 10^6 \\ -60{,}416.66 & -3.625 \times 10^6 & 60{,}416.66 & -3.625 \times 10^6 \\ 3.625 \times 10^6 & 145 \times 10^6 & -3.625 \times 10^6 & 290 \times 10^6 \end{bmatrix} \begin{Bmatrix} u_1 \\ u_2 \\ u_3 \\ u_4 \end{Bmatrix} \quad (3.25)$$

Member 2:

$$\begin{Bmatrix} P_1 \\ P_2 \\ P_3 \\ P_4 \end{Bmatrix} = \begin{bmatrix} 60{,}416.66 & 3.625 \times 10^6 & -60{,}416.66 & 3.625 \times 10^6 \\ 3.625 \times 10^6 & 290 \times 10^6 & -3.625 \times 10^6 & 145 \times 10^6 \\ -60{,}416.66 & -3.625 \times 10^6 & 60{,}416.66 & -3.625 \times 10^6 \\ 3.625 \times 10^6 & 145 \times 10^6 & -3.625 \times 10^6 & 290 \times 10^6 \end{bmatrix} \begin{Bmatrix} u_3 \\ u_4 \\ u_5 \\ u_6 \end{Bmatrix} \tag{3.26}$$

Combining, and using the equivalent nodal loads shown in Figure 3-10, we have

$$\begin{Bmatrix} F_1 \\ F_2 \\ -42{,}120 \\ -1.1232 \times 10^6 \\ F_5 \\ 748{,}800 \end{Bmatrix} = \begin{bmatrix} 60{,}416.66 & 3.625 \times 10^6 & -60{,}416.66 & 3.625 \times 10^6 & 0 & 0 \\ 3.625 \times 10^6 & 290 \times 10^6 & -3.625 \times 10^6 & 145 \times 10^6 & 0 & 0 \\ -60{,}416.66 & -3.625 \times 10^6 & 120{,}833.3 & 0 & -60{,}416.66 & 3.625 \times 10^6 \\ 3.625 \times 10^6 & 145 \times 10^6 & 0 & 580 \times 10^6 & -3.625 \times 10^6 & 145 \times 10^6 \\ 0 & 0 & -60{,}416.66 & -3.625 \times 10^6 & 60{,}416.66 & -3.625 \times 10^6 \\ 0 & 0 & 3.625 \times 10^6 & 145 \times 10^6 & -3.625 \times 10^6 & 290 \times 10^6 \end{bmatrix} \begin{Bmatrix} u_1 \\ u_2 \\ u_3 \\ u_4 \\ u_5 \\ u_6 \end{Bmatrix} \tag{3.27}$$

Since $u_1 = u_2 = u_5 = 0$, the reduced stiffness matrix becomes

$$\begin{Bmatrix} -42{,}120 \\ -1.1232 \times 10^6 \\ 748{,}800 \end{Bmatrix} = \begin{bmatrix} 120{,}833.3 & 0 & 3.625 \times 10^6 \\ 0 & 580 \times 10^6 & 145 \times 10^6 \\ 3.625 \times 10^6 & 145 \times 10^6 & 290 \times 10^6 \end{bmatrix} \begin{Bmatrix} u_3 \\ u_4 \\ u_6 \end{Bmatrix} \tag{3.28}$$

Solving for the displacements we find

$$\begin{Bmatrix} u_3 \\ u_4 \\ u_6 \end{Bmatrix} = \begin{Bmatrix} -.823034\ in \\ -.005890\ rad \\ .015815\ rad \end{Bmatrix} \tag{3.29}$$

Using equations (3.25) and (3.26) to calculate member forces we obtain

Member 1:

$$\begin{Bmatrix} P_1 \\ P_2 \\ P_3 \\ P_4 \end{Bmatrix} = [k]_1 \begin{Bmatrix} 0 \\ 0 \\ -.823034 \\ -.005890 \end{Bmatrix} = \begin{Bmatrix} 28.37\ k \\ 177.5\ ft\text{-}k \\ -28.37\ k \\ 106.3\ ft\text{-}k \end{Bmatrix} \tag{3.30}$$

Member 2:

$$\begin{Bmatrix} P_1 \\ P_2 \\ P_3 \\ P_4 \end{Bmatrix} = [k]_2 \begin{Bmatrix} -.823034 \\ -.005890 \\ 0 \\ .015815 \end{Bmatrix} + \begin{Bmatrix} 42{,}120\ \# \\ 1.1232 \times 10^6\ in\text{-}\# \\ 22{,}880\# \\ -748{,}800\ in\text{-}\# \end{Bmatrix}$$

$$= \begin{Bmatrix} -13{,}747\# \\ -2.398 \times 10^6 in\text{-}\# \\ 13{,}747\# \\ 748{,}802\ in\text{-}\# \end{Bmatrix} + \begin{Bmatrix} 42{,}120 \\ 1.1232 \times 10^6 \\ 22{,}880 \\ -748800 \end{Bmatrix} = \begin{Bmatrix} 28{,}373\ \# \\ -1.275 \times 10^6\ in\text{-}\# \\ 36{,}627\# \\ 2\ in\text{-}\# \end{Bmatrix} \tag{3.31}$$

Figure 3-12 shows free-body diagrams of both members. You should verify that equilibrium of each member is satisfied.

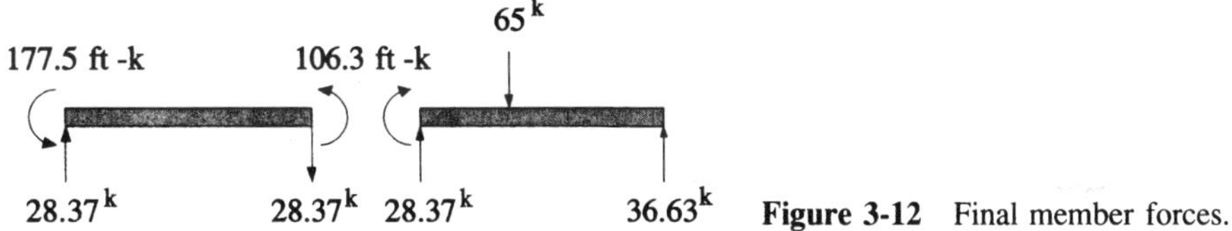

Figure 3-12 Final member forces.

Now that the actual directions of the forces have been sketched in Figure 3-12, we can construct the shearing-force and bending-moment diagrams.

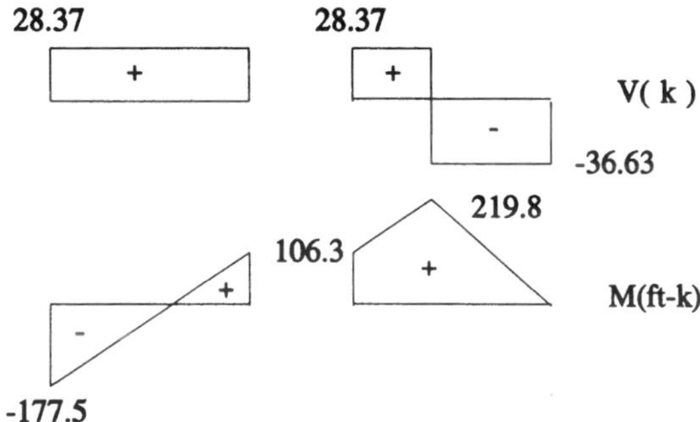

When dealing with non-nodal forces for a frame, we have to account for the case where a frame member is not parallel to either global coordinate direction. This is illustrated in the following example.

Example 3.4

Consider the frame shown in Figure E3-4a.

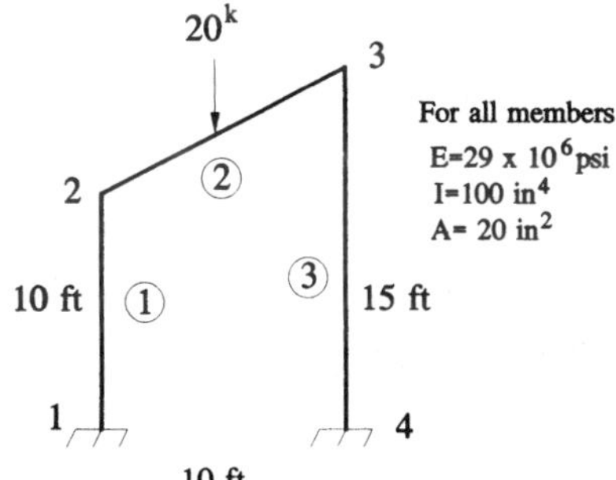

Figure E3-4a

Figure E3-4b shows the fixed end forces acting on member 2 after the 20-kip load has been resolved into components parallel and perpendicular to the member. The equivalent nodal forces are shown in Figure E3-4c. Note that these equivalent forces are the opposite

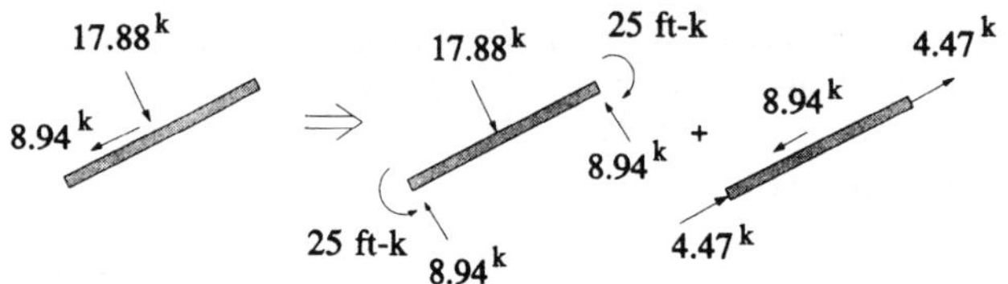

Figure E3-4b Fixed end forces for member 2.

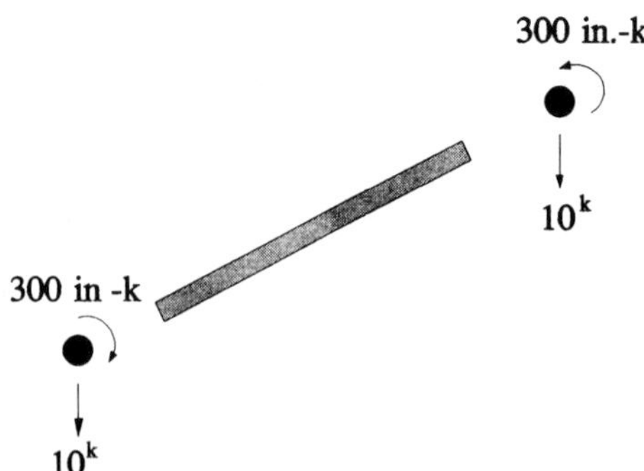

Figure E3-4c Equivalent nodal forces.

of the fixed end forces acting on a beam that spans 10 ft *horizontally*. Of course, to obtain the actual forces on member 2, we will have to add the forces shown in Figure E3-4b to those found by using the calculated displacements.

Using units of inches and kips, equations (3.32), (3.33), and (3.34) represent the elemental stiffness matrices for members 1, 2, and 3, respectively, transformed to the global coordinate system (see equation [3.12]).

$$[k]_1 = \begin{bmatrix} 20.14 & 0 & -1208.33 & -20.14 & 0 & -1208.33 \\ 0 & 4833.33 & 0 & 0 & -4833.33 & 0 \\ -1208.33 & 0 & 96666.66 & 1208.33 & 0 & 48333.33 \\ -20.14 & 0 & 1208.33 & 20.14 & 0 & 1208.33 \\ 0 & -4833.33 & 0 & 0 & 4833.33 & 0 \\ -1208.33 & 0 & 48333.33 & 1208.33 & 0 & 96666.66 \end{bmatrix} \tag{3.32}$$

$$[k]_2 = \begin{bmatrix} 3461.33 & 1723.46 & -432.31 & -3461.33 & -1723.46 & -432.31 \\ 1723.46 & 876.14 & 864.61 & -1723.46 & -876.14 & 864.61 \\ -432.31 & 864.61 & 86461.3 & 432.31 & -864.61 & 43230.65 \\ -3461.33 & -1723.46 & 432.31 & 3461.33 & 1723.46 & 432.31 \\ -1723.46 & -876.14 & -864.61 & 1723.46 & 876.14 & -864.61 \\ -432.31 & 864.61 & 43230.65 & 432.31 & -864.61 & 864611.3 \end{bmatrix} \tag{3.33}$$

$$[k]_3 = \begin{bmatrix} 5.97 & 0 & 537.04 & -5.97 & 0 & 537.04 \\ 0 & 3222.22 & 0 & 0 & -3222.22 & 0 \\ 537.04 & 0 & 64444.45 & -537.04 & 0 & 32222.22 \\ -5.97 & 0 & -537.04 & 5.97 & 0 & -537.04 \\ 0 & -3222.22 & 0 & 0 & 3222.22 & 0 \\ 537.04 & 0 & 32222.22 & -537.04 & 0 & 64444.45 \end{bmatrix} \tag{3.34}$$

Equation (3.35) shows the overall reduced structural stiffness equation obtained after eliminating the rows and columns associated with zero support displacements from the combined equation and applying the equivalent nodal forces.

$$\begin{Bmatrix} 0 \\ -10 \\ -300 \\ 0 \\ -10 \\ 300 \end{Bmatrix} = \begin{bmatrix} 3481.47 & 1723.46 & 776.03 & -3461.33 & -1723.46 & -432.31 \\ 1723.46 & 5709.48 & 864.61 & -1723.46 & -876.14 & 864.61 \\ 776.03 & 864.61 & 183128.0 & 432.31 & -864.61 & 43230.65 \\ -3461.33 & -1723.46 & 432.31 & 3467.30 & 1723.46 & 969.34 \\ -1723.46 & -876.14 & -864.61 & 1723.46 & 4098.36 & -864.61 \\ -432.31 & 864.61 & 43230.65 & 969.34 & -864.61 & 150905.8 \end{bmatrix} \begin{Bmatrix} u_4 \\ u_5 \\ u_6 \\ u_7 \\ u_8 \\ u_9 \end{Bmatrix} \tag{3.35}$$

Solving equation (3.35) for the displacements yields

$$\begin{Bmatrix} u_4 \\ u_5 \\ u_6 \\ u_7 \\ u_8 \\ u_9 \end{Bmatrix} = \begin{Bmatrix} .0747\ in \\ -.0022\ in \\ -.0027\ rad \\ .0746\ in \\ -.0029\ in \\ .0025\ rad \end{Bmatrix} \tag{3.36}$$

Using these displacements the member forces $\{P\}_i = [k_e]_i[\beta]_i\{u\}_i$ become:

Member 1:

$$\begin{Bmatrix} P_1 \\ P_2 \\ P_3 \\ P_4 \\ P_5 \\ P_6 \end{Bmatrix} = \begin{Bmatrix} 10.66\ k \\ -1.79\ k \\ -41.4\ in\text{-}k \\ -10.66\ k \\ 1.79\ k \\ -173.06\ in\text{-}k \end{Bmatrix} \tag{3.37}$$

Member 2:

$$\begin{Bmatrix} P_1 \\ P_2 \\ P_3 \\ P_4 \\ P_5 \\ P_6 \end{Bmatrix} = \begin{Bmatrix} 1.89\ k \\ -.21\ k \\ -126.94\ in\text{-}k \\ -1.89\ k \\ .21\ k \\ 98.88\ in\text{-}k \end{Bmatrix} \tag{3.38}$$

Member 3:

$$\begin{Bmatrix} P_1 \\ P_2 \\ P_3 \\ P_4 \\ P_5 \\ P_6 \end{Bmatrix} = \begin{Bmatrix} 9.34\ k \\ 1.79\ k \\ 201.12\ in\text{-}k \\ -9.34\ k \\ -1.79k \\ 120.58\ in\text{-}k \end{Bmatrix} \tag{3.39}$$

Figure E3-4d shows the combination of the fixed end forces and those from equation (3.38) for member 2. The moments have been converted from in-k to ft-k.

Again, you should verify that equilibrium for each member is satisfied.

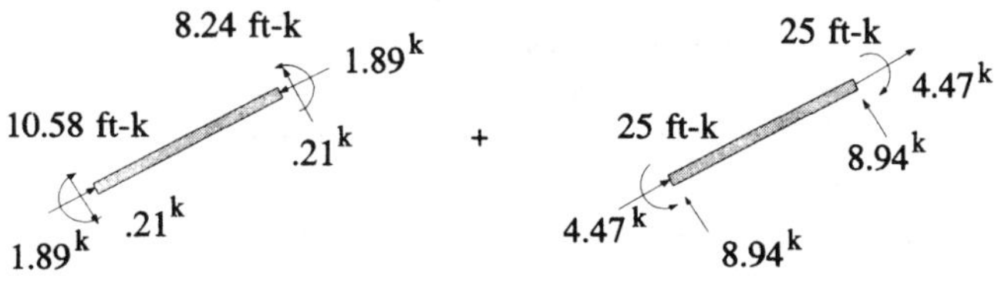

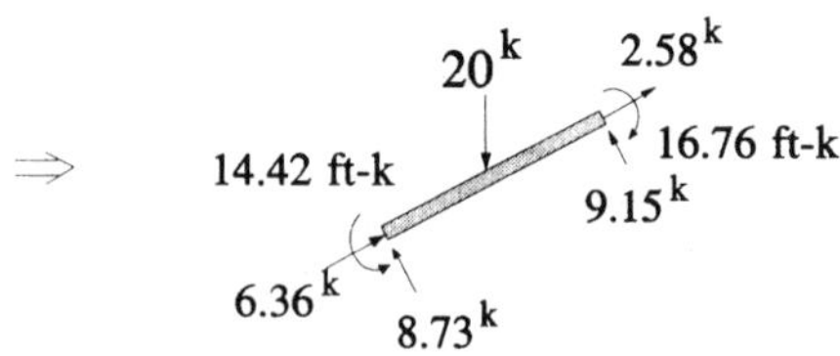

Figure E3-4d Final forces on member 2.

3.7 THERMAL EFFECTS IN BEAMS AND FRAMES

As in the case of the one-dimensional bar, we want to determine the fixed end forces attributable to a temperature change. For the bar we used a constant ΔT over its length and depth. This resulted in only axial fixed end forces. For the beam or frame element we must also consider the development of fixed end moments owing to a temperature variation across the depth of the cross section.

Consider a beam or frame element that is heated along its length and has a linear temperature variation over its depth.

If T_2 and T_1 are the temperatures at the top and bottom of the beam, respectively, then the temperature distribution is given by $T(y) = (T_2 + T_1)/2 + (T_2 - T_1)y/h$, where $T_{av.} = (T_2 + T_1)/2$, y is measured positive up from the centroidal axis of the cross section, and h is the depth of the beam (see Figure 3-13).

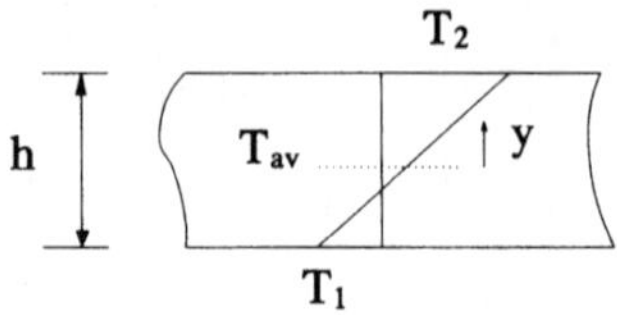

Figure 3-13 Temperature variation over depth of beam.

The strain at any point in the cross section is given by $\varepsilon(y) = \alpha T(y)$, and the corresponding stress (for fixed end conditions) is $\sigma = E\alpha T(y)$. The thermal axial force P_t will be

$$P_t = \int_A \sigma dA = \int_A E\alpha T(y) dA \tag{3.40}$$

If we consider the case of a beam of constant width b, equation (3.40) becomes

$$P_t = \int_{-h/2}^{h/2} E\alpha T(y) b dy = E\alpha A T_{av.} \tag{3.41}$$

The thermal moment will be

$$M_t = \int_{-h/2}^{h/2} \alpha E T(y) b y dy = \frac{\alpha E I (T_2 - T_1)}{h} \tag{3.42}$$

Assuming that T_2 is larger than T_1, the fixed-end moments will act on the beam in clockwise and counterclockwise directions at the left and right ends, respectively. The equivalent nodal forces will act in opposite directions. We now illustrate the solution of the thermal problem for a frame.

Example 3.5

Consider the frame shown in Figure E3-5. This is the same frame used in Example 3.2.

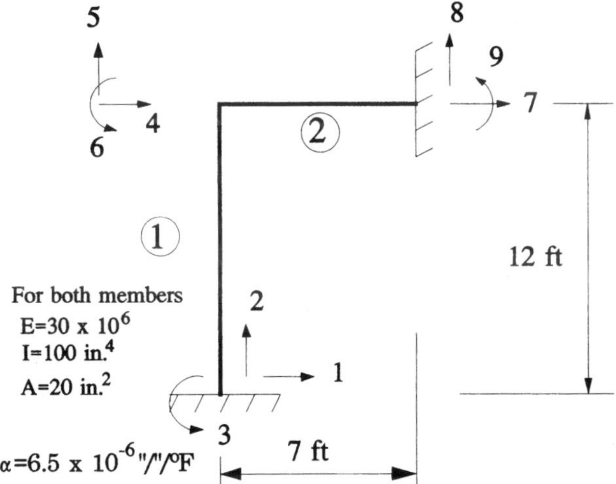

Figure E3-5 Frame for Example 3.5.

Assume member 2 has a temperature variation of +70°F over its depth of 10 in. Equations (3.41) and (3.42) yield:

$$P_t = 6.5 \times 10^{-6}(30 \times 10^6)(20)(35) = 136{,}500\#$$
$$M_t = 6.5 \times 10^{-6}(30 \times 10^6)(100)(70)/10 = 136{,}500''\text{-}\#$$

The equivalent forces at the free node are given by

$$\begin{Bmatrix} -136{,}500\ \# \\ 0 \\ 136{,}500\ in\text{-}\# \end{Bmatrix}$$

Equation (3.21) becomes

$$\begin{Bmatrix} F_4 \\ F_5 \\ F_6 \end{Bmatrix} = \begin{Bmatrix} -136{,}500 \\ 0 \\ 136{,}500 \end{Bmatrix} = \begin{bmatrix} 7.155 \times 10^6 & 0 & 868{,}055 \\ 0 & 4.227 \times 10^6 & 2.551 \times 10^6 \\ 868{,}055 & 2.551 \times 10^6 & 226.19 \times 10^6 \end{bmatrix} \begin{Bmatrix} u_4 \\ u_5 \\ u_6 \end{Bmatrix} \tag{3.43}$$

Solving for the displacements we find $u_4 = -.019160$ in, $u_5 = -.000411$ in, and $u_6 = .000682$ rad. The forces in member 1 become

$$\begin{Bmatrix} P_1 \\ P_2 \\ P_3 \\ P_4 \\ P_5 \\ P_6 \end{Bmatrix} = \begin{Bmatrix} 1710\ \# \\ 360\ \# \\ 11770\ in\text{-}\# \\ -1710\ \# \\ -360\ \# \\ 40170\ in\text{-}\# \end{Bmatrix}$$

The forces in member 2 become

$$\begin{Bmatrix} P_1 \\ P_2 \\ P_3 \\ P_4 \\ P_5 \\ P_6 \end{Bmatrix} = \begin{Bmatrix} -136{,}860\ \# \\ 1710\ \# \\ 96{,}330\ in\text{-}\# \\ 136{,}860\ \# \\ -1710\ \# \\ 47640\ in\text{-}\# \end{Bmatrix} + \begin{Bmatrix} 136{,}500\ \# \\ 0 \\ -136{,}500\ in\text{-}\# \\ -136{,}500\ \# \\ 0 \\ 136{,}500\ in\text{-}\# \end{Bmatrix} = \begin{Bmatrix} -360\ \# \\ 1710\ \# \\ -40{,}170\ in\text{-}\# \\ 360\ \# \\ -1710\ \# \\ 184{,}140\ in\text{-}\# \end{Bmatrix} \tag{3.44}$$

Again, an equilibrium check should be made.

3.8 SUPPORT MOVEMENTS FOR BEAMS AND FRAMES

In Chapter 1 we presented two procedures for analyzing structures with support movements: a numerical method and a matrix partitioning method. Naturally, either of these techniques can be used for any structure; however, in many cases the concept of fixed end forces and equivalent nodal loads is more convenient. This method is illustrated in the following example.

Example 3.6

Support b of the beam shown in Figure E3-6a moves down 0.25 in. Determine the member forces.

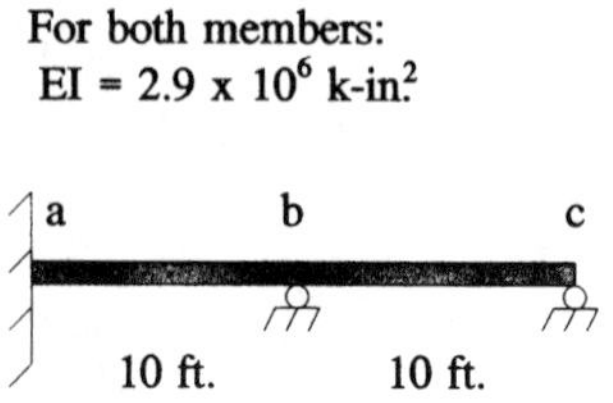

For both members:
EI = 2.9 x 10^6 k-in^2

a b c

10 ft. 10 ft.

Figure E3-6a Beam for Example E3.6.

From the slope-deflection equations, the fixed end moments and shears for a beam with a relative displacement of one end with respect to the other perpendicular to the axis of the member are given by

$$M_F = 6EI\Delta/L^2 \qquad V_F = 12EI\Delta/L^3 \tag{3.45}$$

These fixed end forces and the equivalent nodal loads are shown in Figure E3-6b.

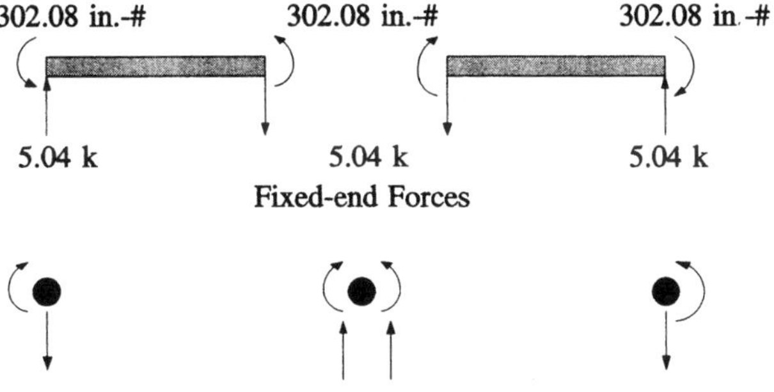

Figure E3-6b Fixed end and equivalent nodal forces.

After applying the equivalent nodal loads and solving for the displacements, we find

$$\begin{Bmatrix} u_1 \\ u_2 \\ u_3 \\ u_4 \\ u_5 \\ u_6 \end{Bmatrix} = \begin{Bmatrix} 0 \\ 0 \\ 0 \\ -.000893\ rad \\ 0 \\ .003571\ rad \end{Bmatrix}$$

from which the member forces shown in Figure E3-6c are obtained.

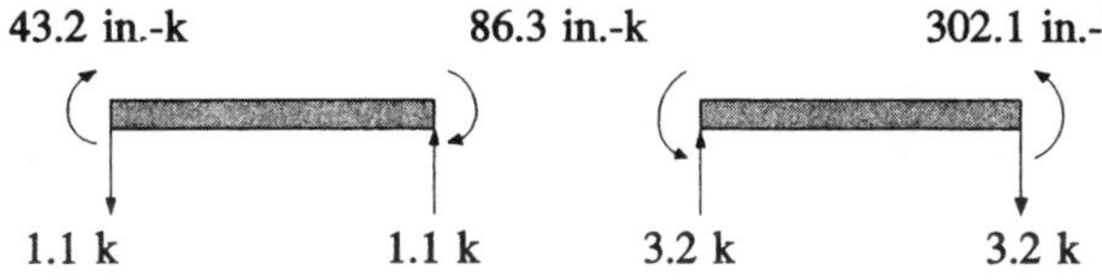

Figure E3-6c Member forces from nodal displacements.

Superposing the fixed end forces shown in Figure E3-6b yields the final member forces shown in Figure E3-6d.

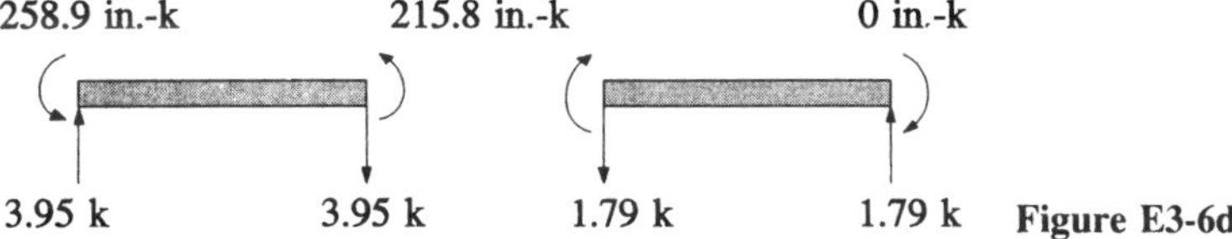

Figure E3-6d

3.9 COMPUTER FORMULATION FOR THE FRAME

We now consider extension of the truss computer formulation to the frame element. Since the beam element is a special case of the frame element with no axial deformation, a frame program can be used to solve beam problems. However, it will not be as efficient as a separately written beam program since the number of degrees of freedom per element is larger (six instead of four). In addition, area properties of the beam elements will have to be entered, and calculations of sines, cosines, and lengths of members will be performed, even though not required for a beam. Memory requirements will also increase, since two coordinates per node must be entered for a frame element, and the size of the global matrix will be three times the number of nodes rather than twice the number of nodes. However, once the frame program is written, it can be quickly reduced to that of a beam program.

For the truss, the global displacement numbers were related to the left and right node numbers of the members. We extend this numbering scheme to the frame. If ML(I) and MR(I) store the left and right node numbers for element I, then the global x and y translations and the z rotation at both ends of the member will be designated as 3*ML(I)-2, 3*ML(I)-1, 3*ML(I), 3*MR(I)-2, 3*MR(I)-1, and 3*MR(I).

As for the case of the truss, after all geometric and material data have been entered we proceed to compute the elements of each member stiffness matrix and place them in the appropriate locations in the global structural stiffness matrix. As in the case of the truss element, we calculate the elemental stiffnesses in terms of global coordinates directly. The following code fragment illustrates the creation of the global stiffness matrix.

Assuming that AK(I)=A(I)*E(I)/L(I), and that S(I) (sine) and C(I) (cosine) for each member have been computed and stored, and letting EIL(I)=E(I)*I(I)/L(I), EIL2(I)= EIL(I)/L(I), and EIL3(I)=EIL2(I)/L(I),

```
FOR I=1 TO NM  [loop on number of members]
EKT(1,1)=AK(I)*C(I)^2+12*EIL3(I)*S(I)^2
EKT(2,1)=AK(I)*S(I)*C(I)-12*EIL3(I)*S(I)*C(I)
... other stiffness terms
IJ(1)=3*ML(I)-2
IJ(2)=3*ML(I)-1
IJ(3)=3*ML(I)
IJ(4)=3*MR(I)-2
IJ(5)=3*MR(I)-1
IJ(6)=3*MR(I)
FOR IR=1 TO 6
FOR IC=1 TO 6
KR=IJ(IR)
KC=IJ(IC)
SK(KR,KC)=SK(KR,KC)+EKT(IR,IC)
NEXT IC: NEXT IR: NEXT I
```

The global stiffness matrix is now filled.

Since we must next construct the reduced stiffness matrix by using the nodal restraint codes (in the case of the frame element we will have KXRES(I), KYRES(I), and KZRES(I) (rotation) entered for each node), we determine the order of the reduced stiffness matrix with the following code fragment:

```
KSUM=0
FOR I=1 TO NN   [loop on number of nodes]
KSUM=KSUM+KXRES(I)+KYRES(I)+KZRES(I)
NEXT I
NKR=3*NN-KSUM  [compute order of reduced stiffness matrix]
```

Making use of the restraint codes, we construct the KEPT() matrix as for the truss. We have a third comparison to make for the frame element.

```
J=0
FOR I=1 TO NN   [loop on nodes]
590 IF(KXRES(I)>0) THEN GOTO 610
600 J=J+1:KEPT(J)=3*I-2
610 IF(KYRES(I)>0) THEN GOTO 630
620 J=J+1:KEPT(J)=3*I-1
630 IF(KZRES(I)>0) THEN GOTO 650
640 J=J+1:KEPT(J)=3*I
650 NEXT I
```

Storing the reduced force and stiffness matrices we have

```
FOR I=1 TO NKR
N=KEPT(I)
FR(I)=F(N)
FOR J=1 TO NKR
M=KEPT(J)
SKR(I,J)=SK(N,M)
NEXT J: NEXT I
```

We are now ready to invert the reduced stiffness matrix. The balance of the procedure, discussed in detail for the truss, is easily extended for the frame. In the case of the frame, however, *all* member forces must be calculated. Naturally, we use

$$\{P\} = [K]\{\delta\} \text{ and } \{\delta\} = [\beta]\{u\}.$$

The reactions are calculated from either the original structural stiffness equation or from the member forces at the nodes corresponding to the reactions.

3.10 SUMMARY

In this chapter we have developed the elemental stiffnesses for both the beam and frame elements. For the frame element it was necessary to generate the beta transformation matrix, which, as in the case of the truss formulation, is used to transform the elemental stiffness matrices to the global coordinate system.

We determined equivalent nodal forces for non-nodal loads applied to beams and frames. Thermal effects in beams and frames were discussed, and support movements were addressed using the concept of equivalent nodal loads. Finally, we outlined the steps necessary to formulate a frame computer program. The basic procedures for the frame program are, of course, identical to those for the bar and truss programs. Only the detailed implementation changes. In the next chapter we will consider the grid element.

PROBLEMS

For the following problems, solve for all nodal displacements and member forces. $E = 29 \times 10^6$ psi for all members.

3.1

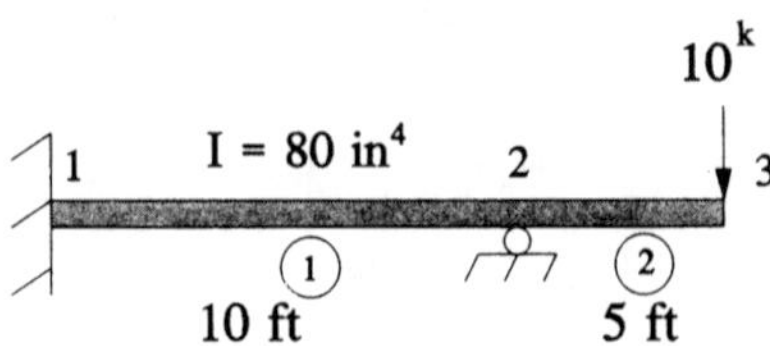

Figure P3-1

3.2

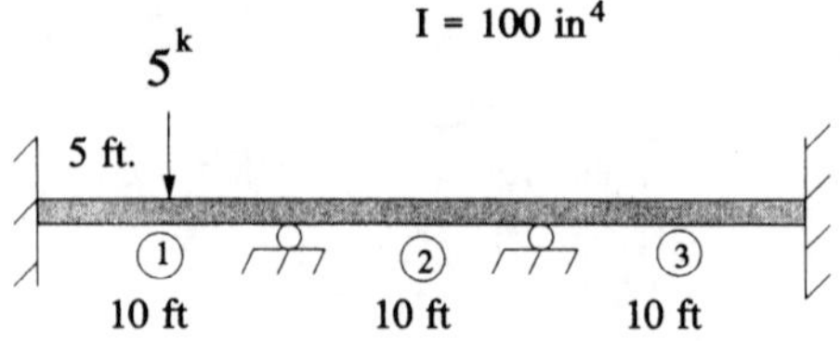

Figure P3-2

3.3

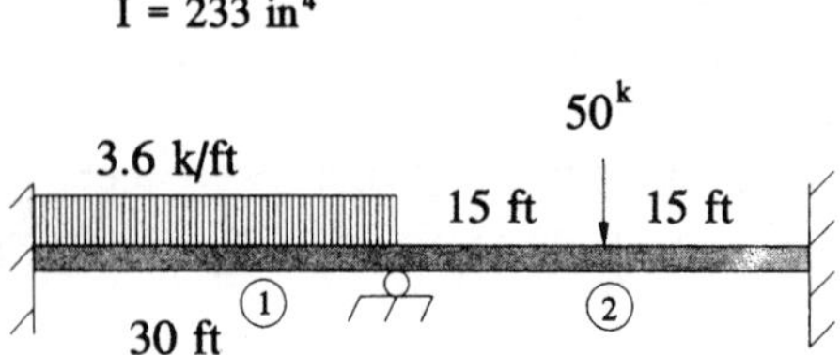

Figure P3-3

3.4

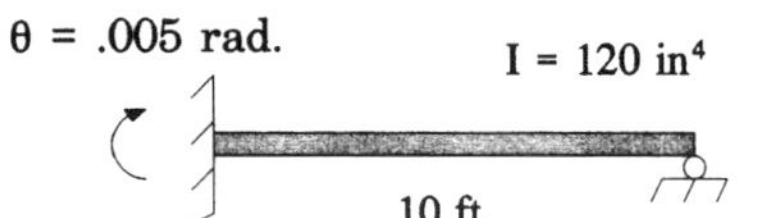

Figure P3-4

3.5

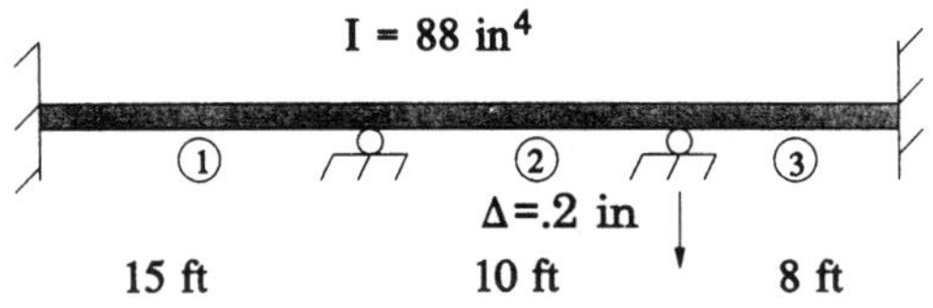

Figure P3-5

3.6

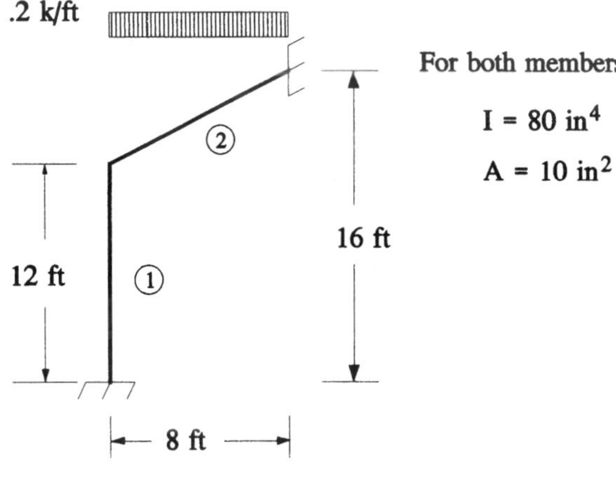

Figure P3-6

3.7

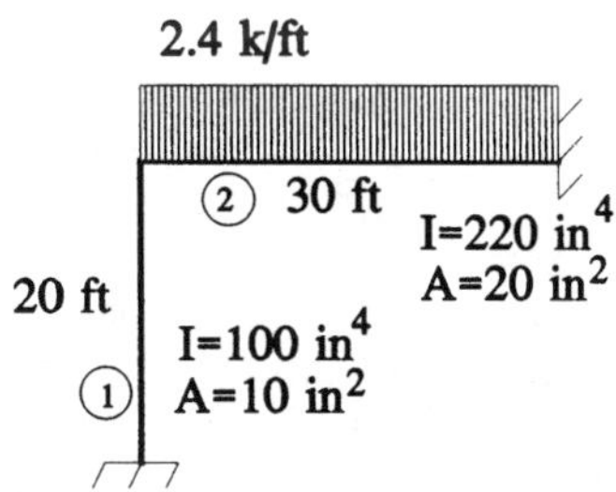

Figure P3-7

3.8

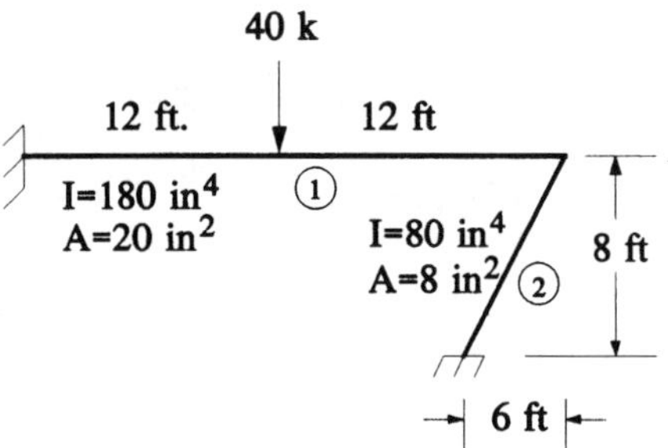

Figure P3-8

3.9 The beam shown has a rectangular cross section with a depth of 8 in. Element 1 has a temperature gradient that varies linearly from 120°F at the top of the beam to 40°F at the bottom. Find all member forces.

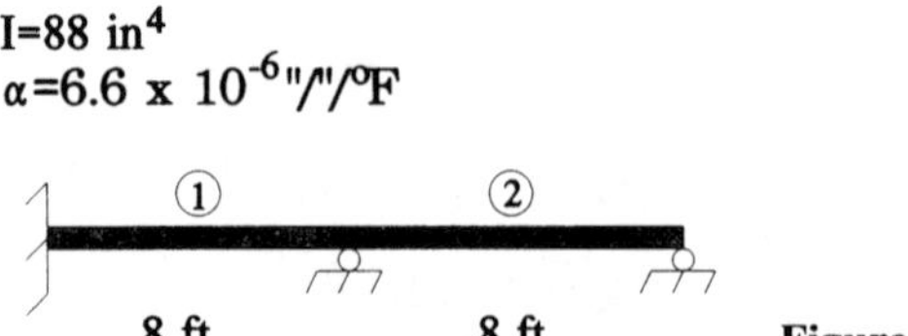

Figure P3-9

3.10 Member 1 in the frame shown in Figure 3.10 has a linear temperature gradient. The temperatures at the top and bottom of the beam are 100°F and 50°F, respectively. The top side of member 1 is on the left of the member. Find all member forces.

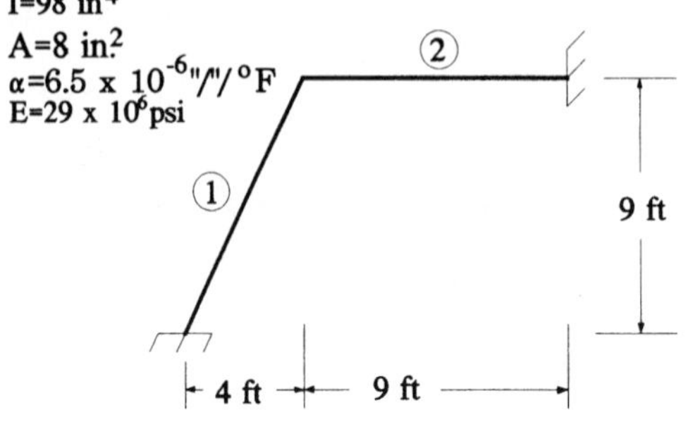

Figure P3-10

3.11 Using the code fragments included in this chapter as a guide, write a computer program to analyze two-dimensional rigid frames. Consider only nodal forces. The program should have the capability to solve problems with 20 nodes and 25 members. (*Hint*: Modify your two-dimensional truss program.)

3.12 Reduce the program written in problem 3.11 to solve two-dimensional beam problems.

For the following problems, use your computer programs written in problems 3.11 and 3.12 as the basis for solution. $E = 29 \times 10^3$ ksi.

3.13

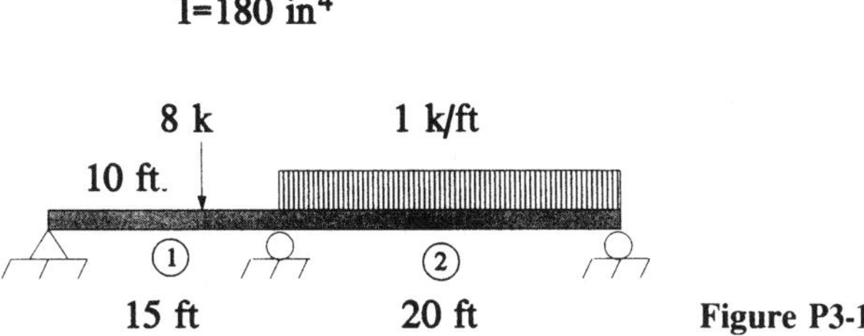

Figure P3-13

3.14

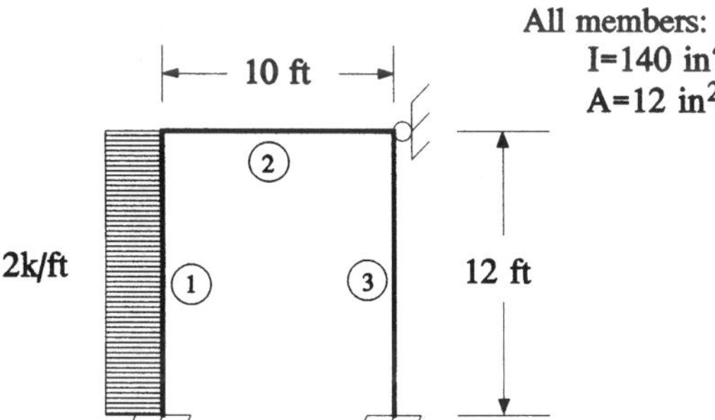

Figure P3-14

3.15

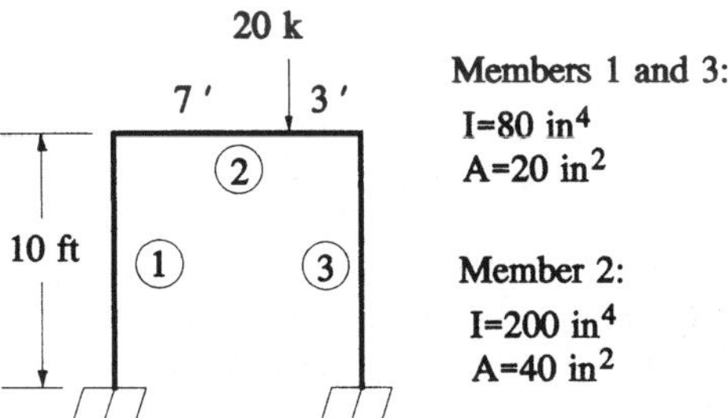

Figure P3-15

3.16

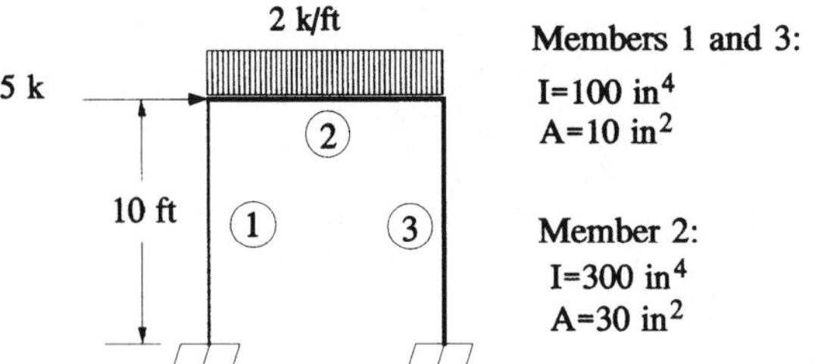

Figure P3-16

3.17

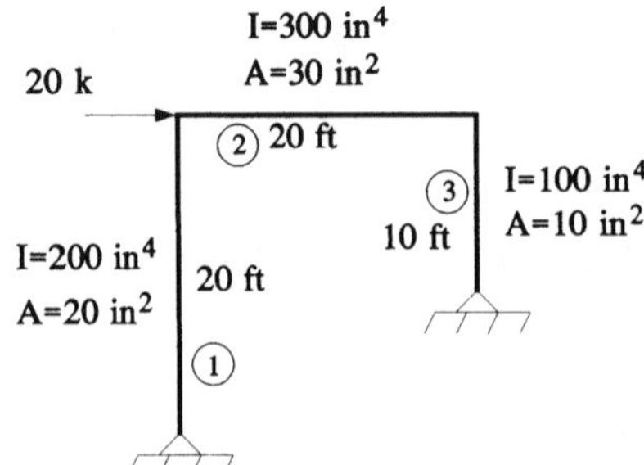

Figure P3-17

3.18

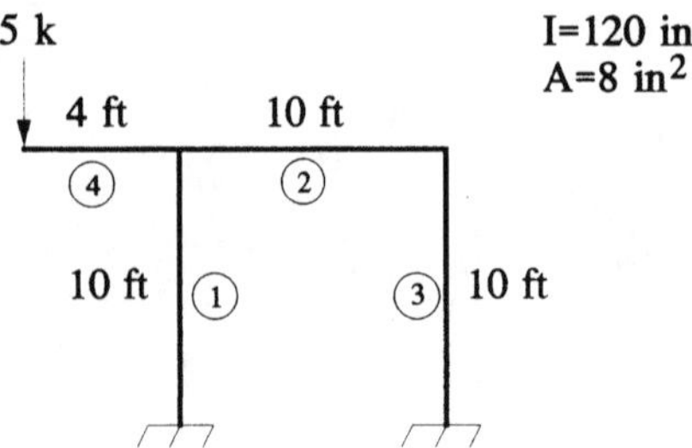

Figure P3-18

3.19

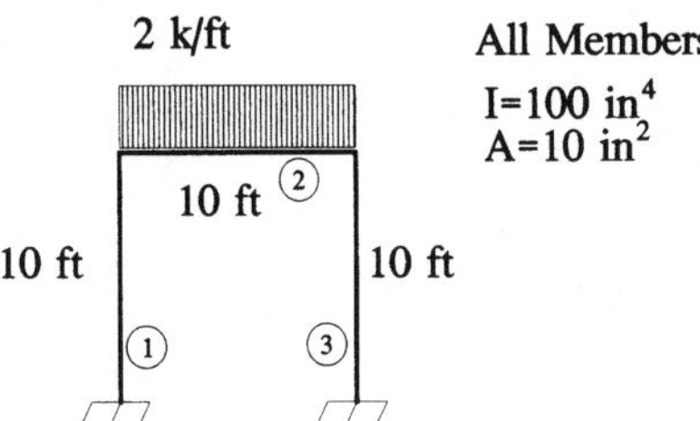

Figure P3-19

3.20 Refer to the two-story frame shown in Figure P3-20.

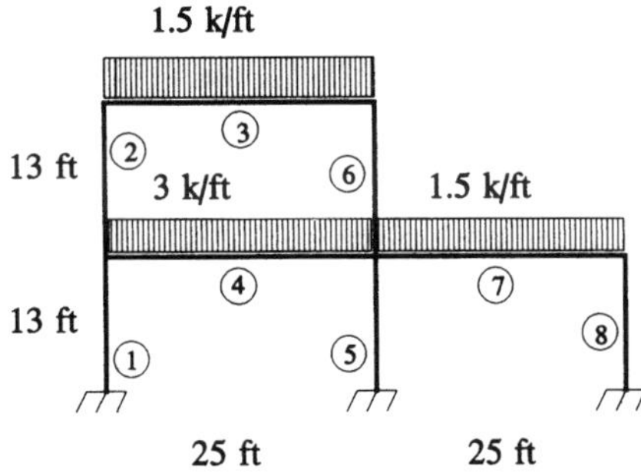

Figure P3-20

Lower story girders: $I = 230$ in^4, $A = 24$ in^2
Upper story girder: $I = 180$ in^4, $A = 14$ in^2
Lower story columns: $I = 103$ in^4, $A = 9.6$ in^2
Upper story columns: $I = 88$ in^4, $A = 7.6$ in^2
Solve for all member forces.

CHAPTER 4

ANALYSIS OF GRIDS

4.1 INTRODUCTION

A grid is a structure that has loads applied perpendicular to its plane. The members are assumed to be rigidly connected at the joints. The floor system shown in Figure 4-1 is an example of a very common grid structure.

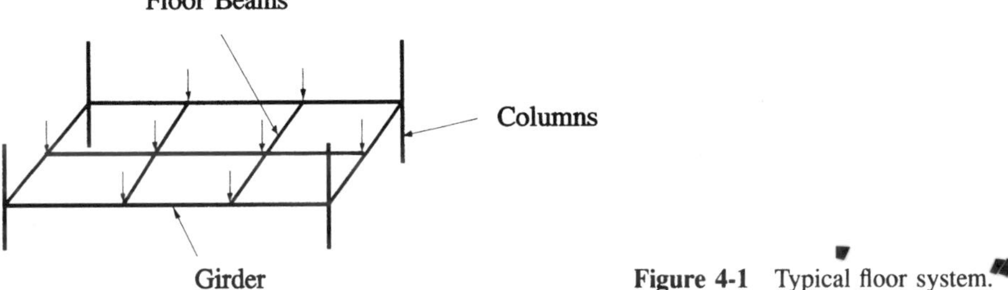

Figure 4-1 Typical floor system.

As in the case of the beam element, we assume that axial deformation is neglected. However, in addition to bending about the horizontal axis of the cross section, the members will also resist the loads by twisting about the axis of the member, thus developing torsional moments. Therefore, at each joint we will have a vertical displacement, a rotation about the horizontal axis of the cross section due to bending, and a rotation about the axis of the member due to torsion. There are three degrees of freedom at each node.

We will use a coordinate system that places the grid in the x-y plane. Vertical loads will therefore act in the z direction and applied nodal moments act in the plane of the grid as shown in Figure 4-2. Figure 4-3 shows the elemental coordinate system we shall use.

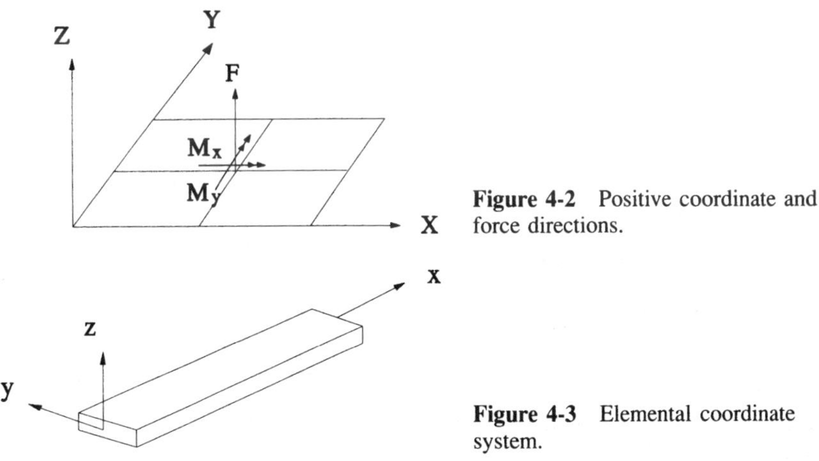

Figure 4-2 Positive coordinate and force directions.

Figure 4-3 Elemental coordinate system.

Bending will occur about the elemental y axis, and twisting occurs about the member x axis. The nodal displacements and forces will be taken as positive when acting in the positive coordinate directions. Naturally, we use the right-hand rule for the direction of the bending and torsional effects. Figure 4-4 shows the positive directions for the elemental forces and displacements.

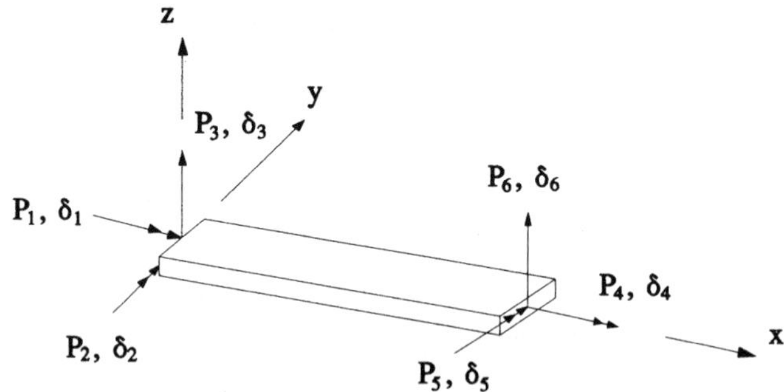

Figure 4-4 Positive elemental forces and displacements.

Note that δ_1, δ_2, δ_4, and δ_5 are rotations and δ_3 and δ_6 are translations.

4.2 DEVELOPMENT OF THE GRID ELEMENTAL STIFFNESS MATRIX

Referring to section 3.2 we can write four of the required six sets of relationships directly from the beam elemental force-displacement equations. For example, referring to the coordinates shown above, with $\delta_3 = 1$ we find:

$$P_2 = -6EI/L^2 = k_{23}, \quad P_3 = 12EI/L^3 = k_{33}, \quad P_5 = -6EI/L^2 = k_{53},$$

$$\text{and } P_6 = -12EI/L^3 = k_{63}.$$

Similarly,

$$\delta_2 = 1, k_{22} = 4EI/L, k_{52} = 2EI/L, k_{32} = -6EI/L^2, k_{62} = 6EI/L^2$$

$$\delta_6 = 1, k_{66} = 12EI/L^3, k_{56} = 6EI/L^2, k_{36} = -12EI/L^3, k_{26} = 6EI/L^2$$

$$\delta_5 = 1, k_{55} = 4EI/L, k_{25} = 2EI/L, k_{65} = 6EI/L^2, k_{35} = -6EI/L^2$$

Recall from your Strength of Materials coursework that the relative angle of rotation of one end of a member of circular cross section with respect to the other when subjected to a torsional moment applied at the centroid of the cross section is given by $\theta = TL/GJ$ as shown in the figure below, where G is the shear modulus and J the polar moment of inertia of the cross section for members with circular cross sections, and the cross sectional torsion constant for other cross sections. Thus, the torsional stiffness is

GJ/L. This elemental stiffness relates the torsional moments P_1 and P_4 to the torsional displacements δ_1 and δ_4.

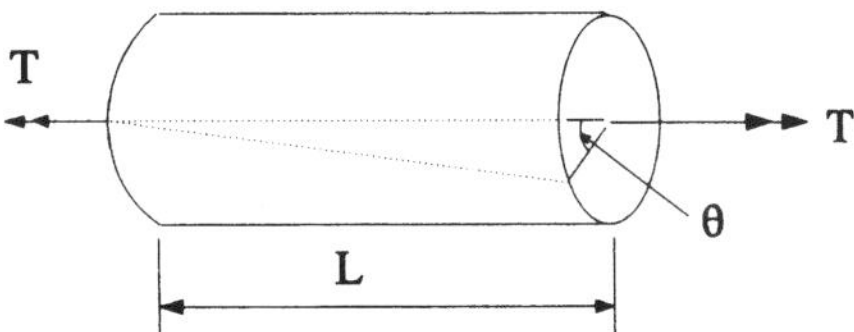

Remember that the circular cross section is the only one for which plane sections remain plane when a torque is applied. That is, there are no displacements parallel to the axis of the bar (perpendicular to the cross section). For cross sections other than circular ones there is some "warping" of the cross section. Some portions of the cross section will have axial displacements in the positive x-axis direction and some portions that will displace in the negative x-axis direction. This warping torsion is generally neglected for most common structures, and we will neglect it in this chapter.

The final elemental force-displacement relationship becomes

$$\begin{Bmatrix} P_1 \\ P_2 \\ P_3 \\ P_4 \\ P_5 \\ P_6 \end{Bmatrix} = \begin{bmatrix} GJ/L & 0 & 0 & -GJ/L & 0 & 0 \\ 0 & 4EI/L & -6EI/L^2 & 0 & 2EI/L & 6EI/L^2 \\ 0 & -6EI/L^2 & 12EI/L^3 & 0 & -6EI/L^2 & -12EI/L^3 \\ -GJ/L & 0 & 0 & GJ/L & 0 & 0 \\ 0 & 2EI/L & -6EI/L^2 & 0 & 4EI/L & 6EI/L^2 \\ 0 & 6EI/L^2 & -12EI/L^3 & 0 & 6EI/L^2 & 12EI/L^3 \end{bmatrix} \begin{Bmatrix} \delta_1 \\ \delta_2 \\ \delta_3 \\ \delta_4 \\ \delta_5 \\ \delta_6 \end{Bmatrix} \quad (4.1)$$

The cross sectional constant for open cross sections made up of thin rectangular shapes is given approximately by $J = \sum_{i=1}^{n} b_i t_i^3/3$ where n is the number of rectangles that make up the cross section. Expressions for J for several common shapes are given in Figure 4-5.

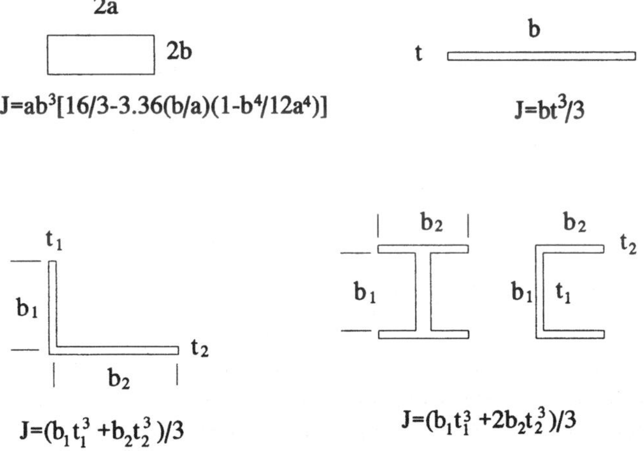

Figure 4-5 Cross-sectional torsional constants.

4.3 COORDINATE TRANSFORMATION

As in the case of the truss and frame elements, we must transform the elemental stiffness matrices with respect to elemental coordinates to the global coordinate system. The global x and y axes will lie in the plane of the structure and are therefore in the same plane as the elemental x and y axes. Elemental and global z axes are parallel to each other.

As shown in Figure 4-6, we must transform the displacements by rotating about the z axis. Defining θ as the angle between the elemental x axis and the global x axis, taken positive counterclockwise, we see that the transformation is identical to that for the frame element (compare Figure 3-7 and Figure 4-4).

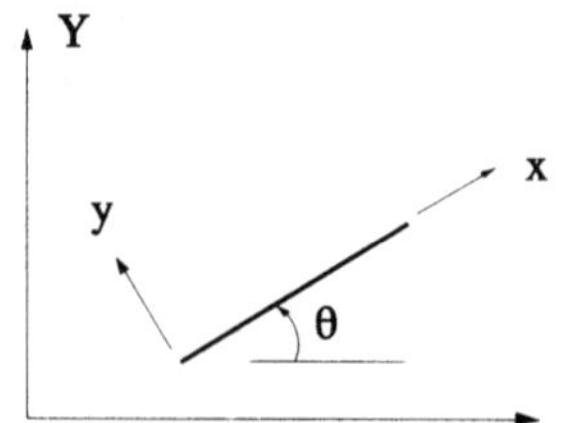

Figure 4-6 Global and elemental coordinate axes.

We can therefore write

$$[\beta] = \begin{bmatrix} \cos\theta & \sin\theta & 0 & 0 & 0 & 0 \\ -\sin\theta & \cos\theta & 0 & 0 & 0 & 0 \\ 0 & 0 & 1 & 0 & 0 & 0 \\ 0 & 0 & 0 & \cos\theta & \sin\theta & 0 \\ 0 & 0 & 0 & -\sin\theta & \cos\theta & 0 \\ 0 & 0 & 0 & 0 & 0 & 1 \end{bmatrix} \tag{4.2}$$

Also,

$$[k_e]_{global} = [\beta]^T [k_e]_{element} [\beta] \tag{4.3}$$

Expanding equation (4.3), we obtain the elemental stiffness matrix in terms of the global coordinate system.

$$\begin{bmatrix} \begin{matrix}(GJ/L)C^2 \\ +(4EI/L)S^2\end{matrix} & \begin{matrix}(GJ/L)SC \\ -(4EI/L)SC\end{matrix} & (6EI/L^2)S & \begin{matrix}(-GJ/L)C^2 \\ +(2EI/L)S^2\end{matrix} & \begin{matrix}(-GJ/L)SC \\ -(2EI/L)SC\end{matrix} & (-6EI/L^2)S \\ & \begin{matrix}(GJ/L)S^2 \\ +(4EI/L)C^2\end{matrix} & -(6EI/L^2)C & \begin{matrix}-(GJ/L)SC \\ -(2EI/L)SC\end{matrix} & \begin{matrix}-(GJ/L)S^2 \\ +(2EI/L)C^2\end{matrix} & (6EI/L^2)C \\ & & 12EI/L^3 & (6EI/L^2)S & (-6EI/L^2)C & (-12EI/L^3) \\ & & & \begin{matrix}(GJ/L)C^2 \\ +(4EI/L)S^2\end{matrix} & \begin{matrix}(GJ/L)SC \\ -(4EI/L)SC\end{matrix} & (-6EI/L^2)S \\ & \textit{symmetric} & & & \begin{matrix}(GJ/L)S^2 \\ +(4EI/L)C^2\end{matrix} & (6EI/L^2)C \\ & & & & & 12EI/L^3 \end{bmatrix} \tag{4.4}$$

4.4 NON-NODAL LOADS

Non-nodal loads are treated in the same way we dealt with them for the beam and frame elements. That is, we will calculate fixed end forces and reverse their directions so as to obtain equivalent nodal forces. After member forces have been found from the calculated displacements, we will add the fixed end forces to obtain the true member forces. This is illustrated in Example 4.2.

4.5 EXAMPLE GRID PROBLEMS

Example 4.1

Consider the grid shown in Figure E4-1a. The member and material properties are indicated to the right of the figure. Note that nodes 2 and 3 are fixed.

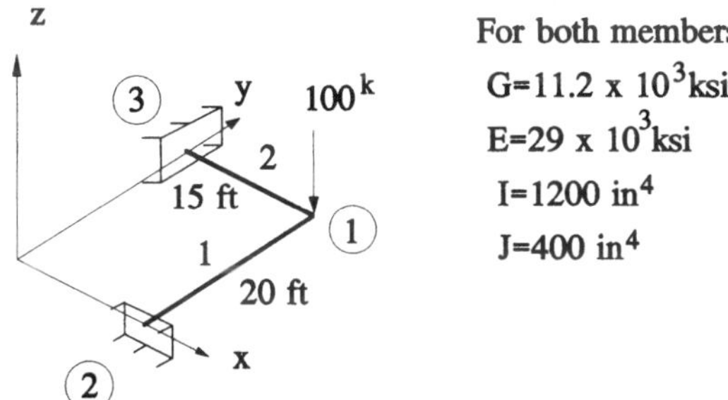

Figure E4-1a Grid for Example 4.1.

For this two-member grid we will assume that node number 1 is the left node of each member. Thus, for member 1, $\theta = 270^0$, $S = -1$ and $C = 0$. For member 2, $\theta = 180^0$, $S = 0$, $C = -1$ (see section 4.3). We next use equation (4.3) to transform each elemental stiffness to the global coordinate system and combine to form the structural stiffness matrix. The individual stiffness matrices using equation (4.4) become:

Member 1:

$$[k]_1 = \begin{bmatrix} 580{,}000 & 0 & -3625 & 290{,}000 & 0 & 3625 \\ - & 18{,}666.7 & 0 & 0 & -18{,}666.7 & 0 \\ - & - & 30.21 & -3625 & 0 & -30.21 \\ - & sym. & - & 580{,}000 & 0 & 3625 \\ - & - & - & - & 18{,}666.7 & 0 \\ - & - & - & - & - & 30.21 \end{bmatrix}$$

Member 2:

$$[k]_2 = \begin{bmatrix} 24{,}888.9 & 0 & 0 & -24{,}888.9 & 0 & 0 \\ - & 773{,}333 & 6444.4 & 0 & 386{,}667 & -6444.4 \\ - & - & 71.61 & 0 & 6444.4 & -71.61 \\ - & sym. & - & 24{,}888.9 & 0 & 0 \\ - & - & - & - & 773{,}333 & -6444.4 \\ - & - & - & - & - & 71.61 \end{bmatrix}$$

Since only u_1, u_2, and u_3 have non-zero values, we need the top left 3 × 3 portion of the structural stiffness matrix. Our reduced equation becomes

$$\begin{Bmatrix} 0 \\ 0 \\ -100^k \end{Bmatrix} = \begin{bmatrix} 604{,}889 & 0 & -3625 \\ 0 & 792{,}000 & 6444 \\ -3625 & 6444 & 101.8 \end{bmatrix} \begin{Bmatrix} u_1 \\ u_2 \\ u_3 \end{Bmatrix} \tag{4.5}$$

Solving equation (4.5) for the unknown displacements, we find;

$$u_1 = -.0217 \text{ rad}, \quad u_2 = .0294 \text{ rad}, \text{ and } u_3 = -3.616 \text{ in}$$

We next find the elemental forces by forming $\{P\} = [k]\{\delta\} = [k][\beta]\{u\}$. Since the right nodes of each member have zero displacements, $\{u\}$ for each member is

$$\{u\} = \begin{Bmatrix} -.0217\ rad \\ .0294\ rad \\ -3.616\ in \\ 0 \\ 0 \\ 0 \end{Bmatrix} \tag{4.6}$$

The elemental forces become

$$\{P\}_1 = [k]_1[\beta]_1\{u\} = \begin{Bmatrix} -549.3 \\ 539.4 \\ -30.7 \\ 549.3 \\ 6824.6 \\ 30.7 \end{Bmatrix} \{P\}_2 = [k]_2[\beta]_2\{u\} = \begin{Bmatrix} 539.4 \\ 549.3 \\ -69.32 \\ -539.41 \\ 11927.7 \\ 69.32 \end{Bmatrix} \tag{4.7}$$

In general,

$$\{P\} = \begin{Bmatrix} GJ/L(Cu_1 + Su_2) - GJ/L(Cu_4 + Su_5) \\ 4EI/L(-Su_1 + Cu_2) + 6EI/L^2(u_6 - u_3) + 2EI/L(-Su_4 + Cu_5) \\ -6EI/L^2(-Su_1 + Cu_2) + 12EI/L^3(u_3 - u_6) - 6EI/L^2(-Su_4 + Cu_5) \\ -P_1 \\ 2EI/L(-Su_1 + Cu_2) + 6EI/L^2(u_6 - u_3) + 4EI/L(-Su_4 + Cu_5) \\ -P_3 \end{Bmatrix}$$

Free-body diagrams of both members are shown in Figure E4-1b with the forces acting in the actual directions.

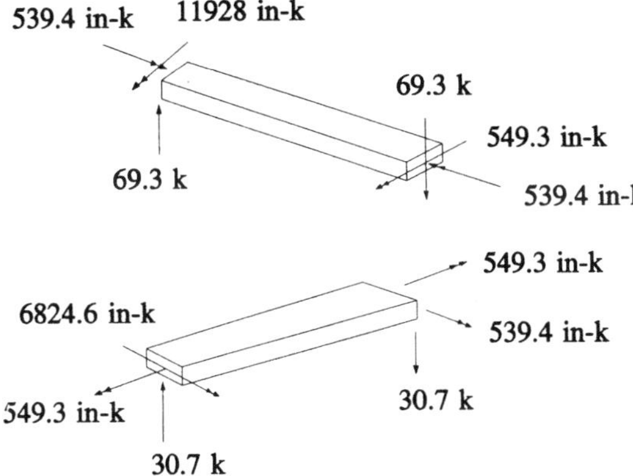

Figure E4-1b Final member forces.

Example 4.2

Consider the grid structure shown in Figure E4-2a.

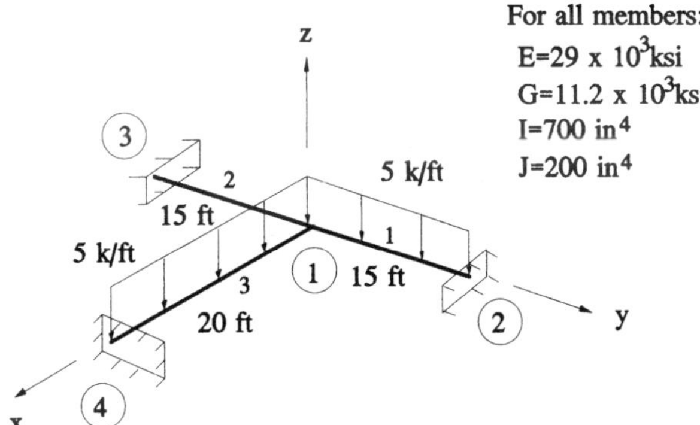

Figure E4-2a Grid for Example 4.2.

Figure E4-2b shows the equivalent nodal forces acting on node 1. These, of course, are the opposite of the fixed end forces.

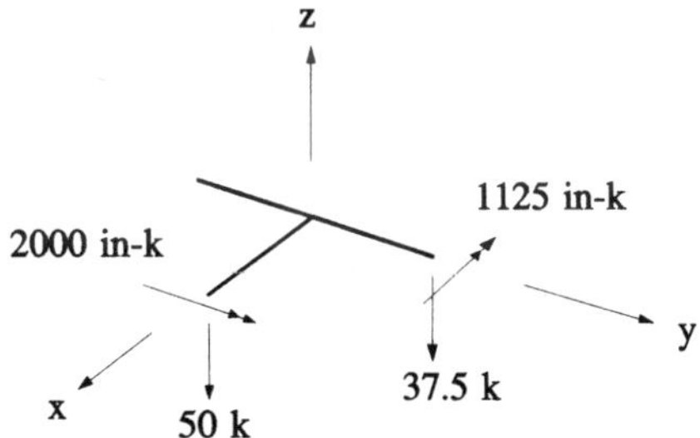

Figure E4-2b Equivalent nodal forces.

Treating node 1 as the left end of each member and noting that node 1 is the only node with non-zero displacements, the top left 3×3 matrix is calculated for each member from equation (4.4). After combining, the reduced structural force-displacement equation becomes

$$\begin{Bmatrix} -1125\ in\text{-}k \\ 2000\ in\text{-}k \\ -87.5\ k \end{Bmatrix} = \begin{bmatrix} 911{,}556 & 0 & 0 \\ 0 & 363{,}222 & -2114.6 \\ 0 & -2114.6 & 101.2 \end{bmatrix} \begin{Bmatrix} u_1 \\ u_2 \\ u_3 \end{Bmatrix} \tag{4.8}$$

Solving for the displacements we find

$$\begin{Bmatrix} u_1 \\ u_2 \\ u_3 \end{Bmatrix} = \begin{Bmatrix} -.001234\ rad \\ .000536\ rad \\ -.853759\ in \end{Bmatrix} \tag{4.9}$$

Figures E4-2c, E4-2d, and E4-2e show the superposition of the fixed end forces to those obtained by using the calculated displacements ($\{P\} = [k][\beta]\{u\}$).

Member 1:

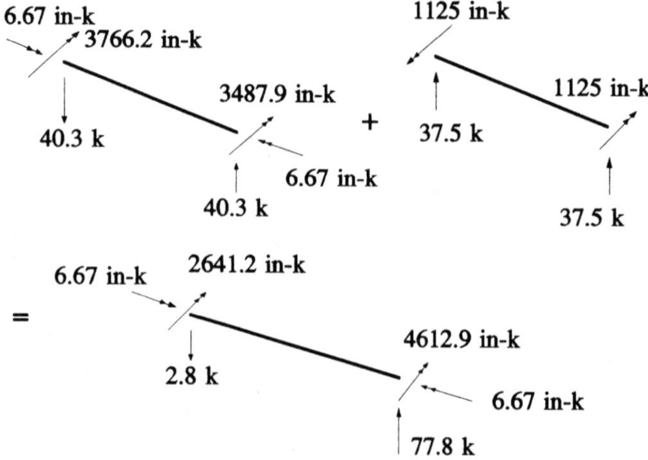

Figure E4-2c Member 1 final forces.

Member 2:

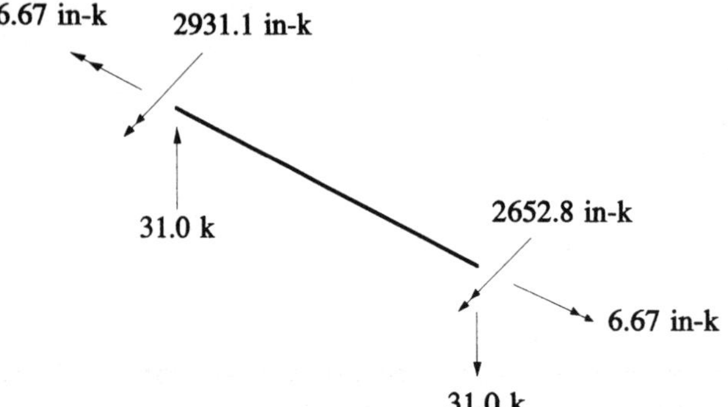

Figure E4-2d Member 2 final forces.

Member 3:

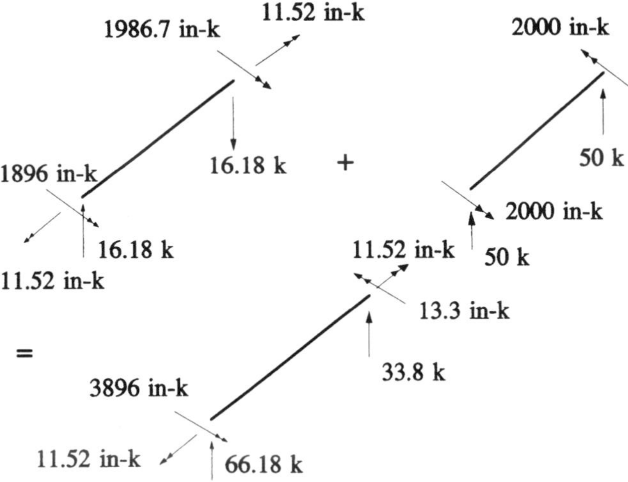

Figure E4-2e Member 3 final forces.

4.6 COMPUTER FORMULATION

Since the $[\beta]$ matrix for the grid element is exactly the same as that for the frame element, and since the displacements at the ends of the grid member are numbered in the same order as those of the frame member, modification of the frame program is relatively simple. Clearly, the elements of the member stiffness matrix must be changed as well as the expressions for member forces. Values for the shear modulus G and the cross sectional torsional constant J must also be entered.

4.7 SUMMARY

The grid element has been addressed in this chapter. It was found that the coordinate transformations required were the same as the frame element transformations. This was due to the selection of the coordinate axes directions and the numbering scheme for the forces and displacements in the elemental coordinate system. Sample problems were presented, including a structure with non-nodal loads applied.

PROBLEMS

For all problems, $E = 29 \times 10^3$ ksi and $G = 11.2 \times 10^3$ ksi.

4.1. A vertical load of -10 k is applied at node 1 of the structure shown in Figure P4-1. Find nodal displacements and member forces. $I = 285\ \text{in}^4$, $J = 0.74\ \text{in}^4$

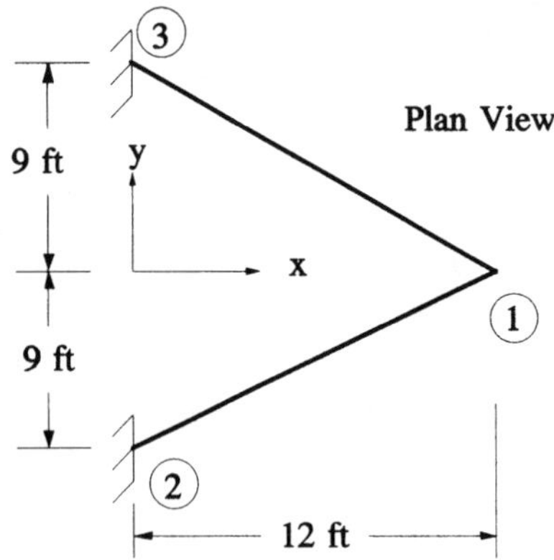

Figure P4-1

4.2. Verify J for the following cross sections:

$$W12 \times 58 : d = 12.19'', \quad t_{\text{web}} = .360'', \quad w_{\text{flange}} = 10.01'', \quad t_{\text{flange}} = .640'', \quad J = 2.10\,\text{in}^4$$

$$\text{Angle } 5 \times 3\text{-}1/2 \times 3/4'' : J = 1.11\ \text{in}^4$$

$$WT9 \times 35.5 : d = 9.235'', t_{\text{stem}} = .495'', w_{\text{flange}} = 7.635'', t_{\text{flange}} = .81'', J = 1.74\ \text{in}^4$$

4.3. Develop a computer program to analyze grids. Modify your frame program to accomplish this task.

Use the grid program of problem 4.3 as the basis for solution of the following problems:

4.4. The members of the structure shown in Figure P4-4 are 8 × 6 × 9/16 structural tubing with $I = 112$ in^4, and $J = 147$ in^4. Find all displacements and member forces.

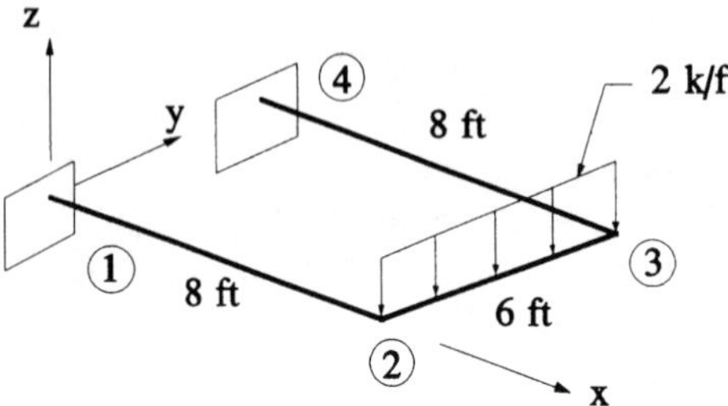

Figure P4-4

4.5. For the structure shown in Figure P4-5, find all member forces. The members are structural tubes 14 × 10 × 5/8 with $I = 728$ in^4 and $J = 885$ in^4. The supports prevent vertical displacements.

4.6. Solve for the displacements and member forces for the cantilevered grid shown in Figure P4-6. The members are 8 × 8 × 5/8 structural tubing with $I = 153$ in^4 and $J = 258$ in^4.

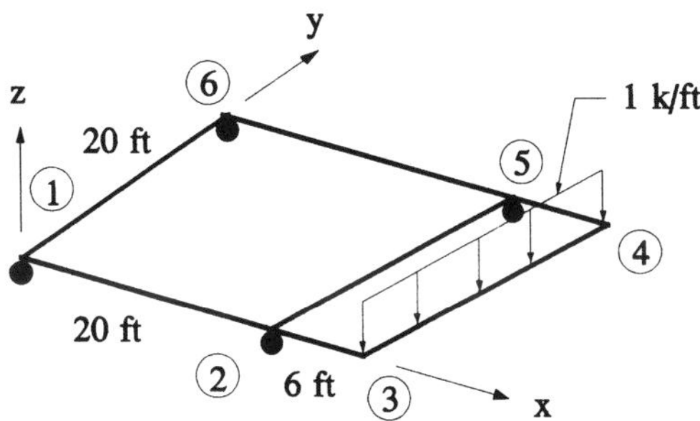

Figure P4-5

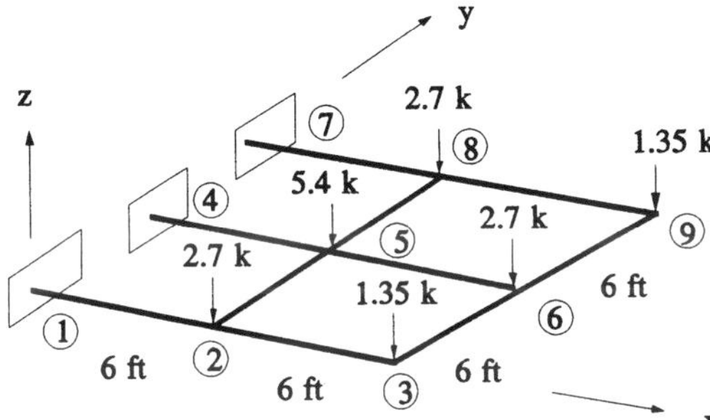

Figure P4-6

4.7. The members of the floor system shown in Figure P4-7 consist of $10 \times 8 \times 5/8$ structural tubing. $I = 266$ in^4 and $J = 367$ in^4 for all members. Find the displacements and member forces. The supports prevent vertical displacement.

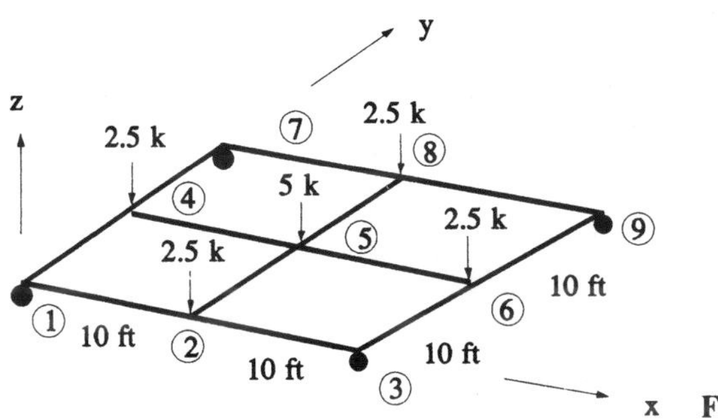

Figure P4-7

4.8. Replace the members in problem 4.7 with W 10×49 having an $I = 272$ in^4 and a $J = 1.39$ in^4. Compare the results with those of problem 4.7.

4.9. Solve problem 4.8 with complete constraint at the corner nodes. Compare with the results of problem 4.8.

CHAPTER 5

THREE-DIMENSIONAL TRUSSES

5.1 INTRODUCTION

Many structures lend themselves to modeling as a combination of two-dimensional structures. For example, the bridge truss shown in Figure 5-1 could be analyzed for vertical loads by considering each side of the bridge as a two-dimensional truss. The loads are delivered to the joints of the truss by floor beams and floor girders.

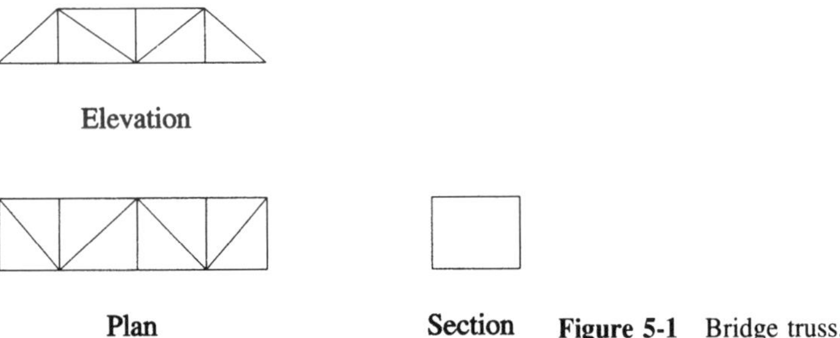

Figure 5-1 Bridge truss.

Other structures, such as domes, aircraft structures, and guyed towers are inherently three-dimensional. They depend on their three-dimensional geometry to support and transfer loads that are applied to them.

For a three-dimensional truss, we make the same assumptions as we made in Chapter 2 for the two-dimensional truss. That is, the members are straight, loads are applied only to the joints, and the members are connected at the joints with frictionless ball joints. These assumptions result in each member acting as a two-force member with either tensile or compressive forces acting along its axis.

The primary difference between the two-dimensional and three-dimensional truss elements is the number of degrees of freedom per node. Since there are three translations possible at each node for the three-dimensional case, each member has six degrees of freedom. We therefore need to expand our elemental stiffness matrix to a 6×6 size. The $[\beta]$ coordinate transformation matrix, which relates the elemental and global displacements, will also increase in size to a 6×6 matrix. The process for formulating and solving a three-dimensional truss problem remains conceptually the same as for the elements previously considered.

5.2 ELEMENTAL STIFFNESS MATRIX

As for all elements discussed earlier, the elemental stiffness matrix will first be expressed in terms of the elemental coordinate system. The transformation to the global coordinate system will then be performed.

Figure 5-2 shows the positive directions of the member axes and forces that we will use for the elemental stiffness.

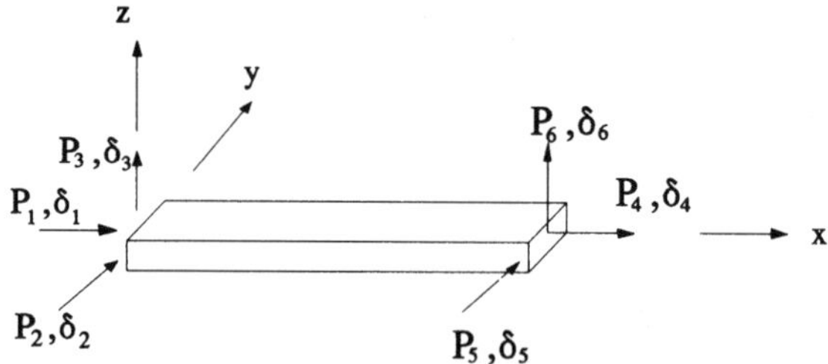

Figure 5-2 Elemental forces and displacements.

Using the axis of the member as the elemental x direction as shown in Figure 5-2, and assuming the positive direction is from the left node toward the right node, we find, by expanding the two-dimensional truss stiffness to three dimensions by adding two zero rows and columns corresponding to F_3, F_6, δ_3, and δ_6 to represent displacements and forces in the third direction, that

$$[k_e]_{element} = EA/L \begin{bmatrix} 1 & 0 & 0 & -1 & 0 & 0 \\ 0 & 0 & 0 & 0 & 0 & 0 \\ 0 & 0 & 0 & 0 & 0 & 0 \\ -1 & 0 & 0 & 1 & 0 & 0 \\ 0 & 0 & 0 & 0 & 0 & 0 \\ 0 & 0 & 0 & 0 & 0 & 0 \end{bmatrix} \tag{5.1}$$

5.3 COORDINATE TRANSFORMATION

For the two-dimensional truss element we found that the $[\beta]$ transformation matrix was given by equation (5.2).

$$[\beta] = \begin{bmatrix} \cos\theta & \sin\theta & 0 & 0 \\ -\sin\theta & \cos\theta & 0 & 0 \\ 0 & 0 & \cos\theta & \sin\theta \\ 0 & 0 & -\sin\theta & \cos\theta \end{bmatrix} \tag{5.2}$$

The 2×2 submatrices consisting of sines and cosines are matrices of direction cosines. That is, each element of this matrix represents the cosine of the angle between the local (elemental) and global coordinate axes. We designate these elements as l_{ij}, where i and j correspond to the elemental displacement δ and the global displacement u, respectively. For example, in the two-dimensional case (refer to Figure 5-3),

l_{11} = cosine of the angle between δ_1 and $u_1 = \cos\theta$
l_{12} = cosine of the angle between δ_1 and $u_2 = \sin\theta$
l_{21} = cosine of the angle between δ_2 and $u_1 = -\sin\theta$
l_{22} = cosine of the angle between δ_2 and $u_2 = \cos\theta$

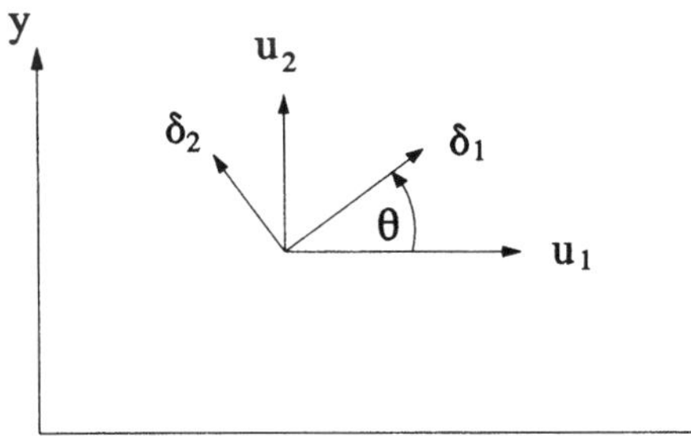

Figure 5-3 Elemental and global displacements.

Expanding to three dimensions, we have for each 3 × 3 submatrix,

$$[L] = \begin{bmatrix} l_{11} & l_{12} & l_{13} \\ l_{21} & l_{22} & l_{23} \\ l_{31} & l_{32} & l_{33} \end{bmatrix} \tag{5.3}$$

Since the transformations required are the same at each end of the member, the required 6 × 6 $[\beta]$ matrix becomes

$$[\beta] = \begin{bmatrix} [L] & [0] \\ [0] & [L] \end{bmatrix} = \begin{bmatrix} l_{11} & l_{12} & l_{13} & 0 & 0 & 0 \\ l_{21} & l_{22} & l_{23} & 0 & 0 & 0 \\ l_{31} & l_{32} & l_{33} & 0 & 0 & 0 \\ 0 & 0 & 0 & l_{11} & l_{12} & l_{13} \\ 0 & 0 & 0 & l_{21} & l_{22} & l_{23} \\ 0 & 0 & 0 & l_{31} & l_{32} & l_{33} \end{bmatrix} \tag{5.4}$$

As previously derived,

$$[k_e]_{global} = [\beta]^T [k_e]_{element} [\beta] \tag{5.5}$$

Using equations (5.1) and (5.4) in equation (5.5) we find

$$[k_e]_{global} = EA/L \begin{bmatrix} l_{11}^2 & l_{11}l_{12} & l_{11}l_{13} & -l_{11}^2 & -l_{11}l_{12} & -l_{11}l_{13} \\ l_{11}l_{12} & l_{12}^2 & l_{12}l_{13} & -l_{11}l_{12} & -l_{12}^2 & -l_{12}l_{13} \\ l_{13}l_{11} & l_{13}l_{12} & l_{13}^2 & -l_{11}l_{13} & -l_{12}l_{13} & -l_{13}^2 \\ -l_{11}^2 & -l_{11}l_{12} & -l_{11}l_{13} & l_{11}^2 & l_{11}l_{12} & l_{11}l_{13} \\ -l_{11}l_{12} & -l_{12}^2 & -l_{12}l_{13} & l_{11}l_{12} & l_{12}^2 & l_{12}l_{13} \\ -l_{13}l_{11} & -l_{13}l_{12} & -l_{13}^2 & l_{11}l_{13} & l_{13}l_{12} & l_{13}^2 \end{bmatrix} \tag{5.6}$$

Note that in the above equation only the direction cosines l_{11}, l_{12}, and l_{13} appear. These represent the cosines of the angles between the δ_1 axis (along the length of the member) and the u_1, u_2, and u_3 directions. We next determine expressions for these three direction cosines.

Consider the vector $\overrightarrow{\mathbf{AB}}$ directed along the member axis as shown in Figure 5-4. This vector could represent either a force or displacement directed along the axis of the member. The global coordinate axes and a unit vector $\vec{\lambda}$ in the AB direction are also shown.

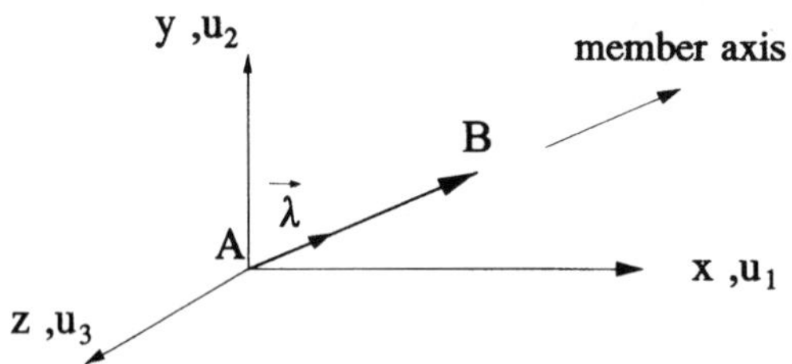

Figure 5-4 Vector along member axis.

From vector algebra we can write

$$\overrightarrow{\mathbf{AB}} = AB\vec{\boldsymbol{\lambda}} = AB(\lambda_x\vec{i} + \lambda_y\vec{j} + \lambda_z\vec{k}) \tag{5.7}$$

where λ_x, λ_y, and λ_z are l_{11}, l_{12}, and l_{13} (the direction cosines of the angles between the vector and the coordinate axes). Now,

$$\overrightarrow{\mathbf{AB}} = (x_B - x_A)\vec{i} + (y_B - y_A)\vec{j} + (z_B - z_A)\vec{k} = AB\vec{\lambda} \tag{5.8}$$

where the difference in coordinates between points A and B have been used to compute the components of the vector $\overrightarrow{\mathbf{AB}}$, and the magnitude of $\overrightarrow{\mathbf{AB}}$ is designated as AB (the length of the vector).

$$AB = \sqrt{(x_b - x_A)^2 + (y_B - y_A)^2 + (z_B - z_A)^2} \tag{5.9}$$

From equation (5.8) we can see that

$$\vec{\lambda} = (x_B - x_A)\vec{i}/AB + (y_B - y_A)\vec{j}/AB + (z_B - z_A)\vec{k}/AB \tag{5.10}$$

Identifying terms with those of equation (5.7) we have

$$\begin{aligned} l_{11} &= \lambda_x = (x_B - x_A)/AB \\ l_{12} &= \lambda_y = (y_B - y_A)/AB \\ l_{13} &= \lambda_z = (z_B - z_A)/AB \end{aligned} \tag{5.11}$$

We see that by calculating differences in coordinates of the two points and the length of the vector, we easily determine the direction cosines needed for our coordinate transformation.

5.4 EXAMPLES OF THREE-DIMENSIONAL TRUSS PROBLEMS

Example 5-1

Consider the three-dimensional truss shown in Figure E5-1.

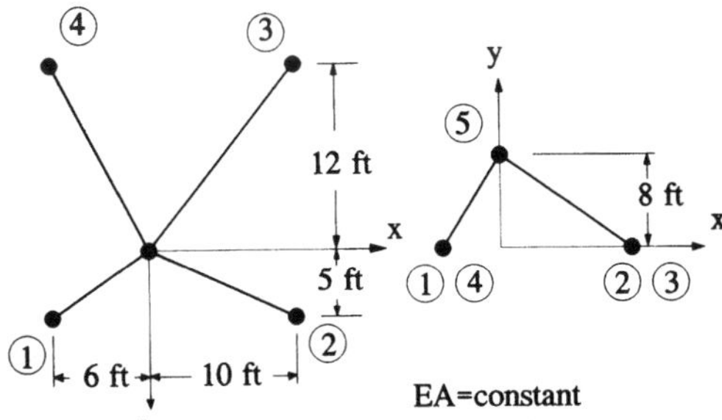

Figure E5-1 Truss for Example 5-1.

Assume that all supports (nodes 1 through 4) are ball joints and therefore prevent any translations. We first compute the direction cosines of each member, treating node number 5 as the left end of each.

Member 5-1:

$$\sqrt{6^2 + 8^2 + 5^2} = 11.18 \text{ ft}$$

$$l_{11} = -6/11.18 = -.5366 \quad l_{12} = -8/11.18 = -.7155 \quad l_{13} = 5/11.18 = .4472$$

Member 5-2:

$$\sqrt{10^2 + 8^2 + 5^2} = 13.75 \text{ ft}$$

$$l_{11} = 10/13.75 = .7274 \quad l_{12} = -8/13.75 = -.5819 \quad l_{13} = 5/13.75 = .3637$$

Member 5-3:

$$\sqrt{10^2 + 8^2 + 12^2} = 17.55 \text{ ft}$$

$$l_{11} = 10/17.55 = .5698 \quad l_{12} = -8/17.55 = -.4558 \quad l_{13} = -12/17.55 = -.6838$$

Member 5-4:

$$\sqrt{6^2 + 8^2 + 12^2} = 15.62 \text{ ft}$$

$$l_{11} = -6/15.62 = -.3841 \quad l_{12} = -8/15.62 = -.5121 \quad l_{13} = -12/15.62 = -.7682$$

Note that since node 5 is the left end of each member and is also the only node with non-zero displacements, only the upper left 3×3 matrix for each member will contribute to the global structural stiffness matrix. Using equation (5.6), the top left 3×3 for each of the members becomes

Member 5-1:

$$EA \begin{bmatrix} .02575 & .03434 & -.02147 \\ .03434 & .04579 & -.02862 \\ -.02147 & -.02862 & .01789 \end{bmatrix} \tag{5.12}$$

Member 5-2:

$$EA\begin{bmatrix} .03849 & -.03079 & .01924 \\ -.03079 & .02463 & -.01539 \\ .01924 & -.01539 & .00962 \end{bmatrix} \tag{5.13}$$

Member 5-3:

$$EA\begin{bmatrix} .01850 & -.01480 & -.02222 \\ -.01480 & .01184 & .01776 \\ -.02222 & .01776 & .02664 \end{bmatrix} \tag{5.14}$$

Member 5-4:

$$EA\begin{bmatrix} .00944 & .01260 & .01890 \\ .01260 & .01680 & .02520 \\ .01890 & .02520 & .03780 \end{bmatrix} \tag{5.15}$$

The x, y, and z translations at each joint are labeled in a manner similar to the two-dimensional truss. These displacements are $3 \times NN - 2$, $3 \times NN - 1$, and $3 \times NN$, where NN is the node number. Thus for node 5 the displacement subscripts are 13, 14, and 15.

Combining the above matrices we have

$$\begin{Bmatrix} F_{13} \\ F_{14} \\ F_{15} \end{Bmatrix} = EA\begin{bmatrix} .09218 & .00135 & -.00553 \\ .00135 & .09906 & -.00105 \\ -.00553 & -.00105 & .09195 \end{bmatrix}\begin{Bmatrix} u_{13} \\ u_{14} \\ u_{15} \end{Bmatrix} \tag{5.16}$$

Let us assume that there is a force applied at node 5, which has components $F_x = 1$ kip, $F_y = 2$ kips, $F_z = 3$ kips. Using $E = 30 \times 10^3$ ksi and $A = 4$ in^2 for all members, solving equation (5.16) for the unknown displacements yields $u_{13} = u_x = .0001047$ ft, $u_{14} = u_y = .0001698$ ft, and $u_{15} = u_z = .0002802$ ft.

We next calculate the member forces using $\{P\} = [k]_{element}[\beta]\{u\}$.

As in the case of the two-dimensional truss, we need to compute only one force. In the three-dimensional case this is P_4 (see Figure 5-2). In general,

$$P_4 = EA/L\,[l_{11}(u_4 - u_1) + l_{12}(u_5 - u_2) + l_{13}(u_6 - u_3)] \tag{5.17}$$

Using equation (5.17) we find,

$$P_{5\text{-}1} = .5626\ k,\ P_{5\text{-}2} = -.6918\ k,\ P_{5\text{-}3} = 1.4315\ k,\ \text{and}\ P_{5\text{-}4} = 2.6310\ k.$$

We now perform an equilibrium check in the x-direction at node 5. For each member the component in the x-direction is Pl_{11}. Thus,

$$\Sigma F_x = .5625(-.5366) - .6918(.7274) + 1.4315(.5698) + 2.6310(-.3841) + 1^k = 0\text{(check)}$$

You should perform equilibrium checks in the y and z directions as an exercise.

Example 5-2

In addition to the applied load at node 1 of the truss shown in Figure E5-2, members 1-3 and 1-4 are subjected to a temperature rise of 50°F. Using $E = 29 \times 10^3$ ksi and $\alpha = 6.5 \times 10^{-6''}/''/^\circ$F, find the final bar forces. The area of members 1-3 and 1-4 is 2 in^2. Member 1-2 has an area of 5 in^2. Nodes 2, 3, and 4 are ball-joint supports.

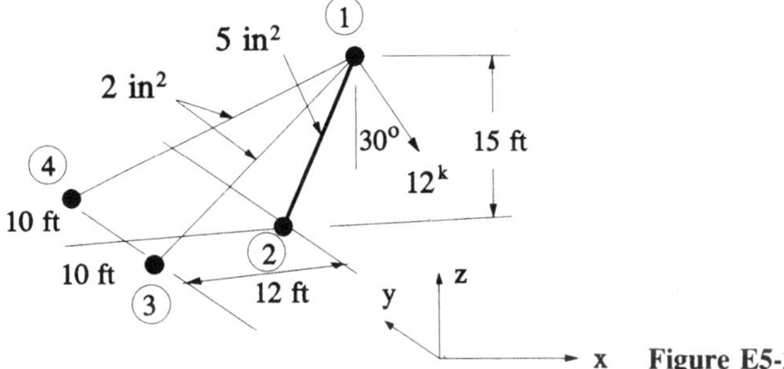

Figure E5-2

Letting node 1 be the left end of each member, the direction cosines are found to be the following:

Member 1-2: $l_{11} = -.3714$, $l_{12} = 0$, $l_{13} = -.9285$
Member 1-3: $l_{11} = -.7066$, $l_{12} = -.3925$, $l_{13} = -.5888$
Member 1-4: $l_{11} = -.7066$, $l_{12} = .3925$, $l_{13} = -.5888$

As in the previous example, we calculate the upper left 3×3 matrix for each member. After combining, we obtain the following reduced structural stiffness equation:

$$\begin{Bmatrix} F_1 \\ F_2 \\ F_3 \end{Bmatrix} = E \begin{bmatrix} .1211 & 0 & .1721 \\ 0 & .0242 & 0 \\ .1721 & 0 & .3214 \end{bmatrix} \begin{Bmatrix} u_1 \\ u_2 \\ u_3 \end{Bmatrix} \tag{5.18}$$

We need to find the equivalent nodal loads due to the temperature changes in members 1-3 and 1-4. Each member fixed end force is given by $EA\alpha(\Delta t) = 18.85$ kips (compression). Reversing these forces and using the direction cosines for each member to calculate the components in the global coordinate directions, we find

$$\begin{aligned} F_1 &= 12^k \sin 30^o + .7066(18.850)^k + .7066(18.850)^k = 32.64^k \\ F_2 &= 0 + .3925(18.850)^k - .3925(18.850)^k = 0 \\ F_3 &= -12 \cos 30^o + .5888(18.850)^k + .5888(18.850)^k = 11.81^k \end{aligned} \tag{5.19}$$

Using equation (5.18) in equation (5.19) and solving for the displacements, we have

$$\begin{Bmatrix} u_1 \\ u_2 \\ u_3 \end{Bmatrix} = \begin{Bmatrix} .03136 \\ 0 \\ -.01553 \end{Bmatrix} \text{ ft} \tag{5.20}$$

Remembering that we must add the fixed end forces to those computed using these displacements, we find

$$P_{1\text{-}2} = -24.83\ k$$
$$P_{1\text{-}3} = 29.57 - 18.85 = 10.72\ k$$
$$P_{1\text{-}4} = 29.57 - 18.85 = 10.72\ k$$

Again, checking equilibrium in the x-direction at node 1 we have:

$$\Sigma F_x = -24.83(-.3714) + 10.72(-.7066) + 10.72(-.7066) + 12 \sin 30^\circ = .07 \approx 0$$

Equilibrium checks with some round-off error. As an exercise you should check equilibrium in the other coordinate directions.

5.5 COMPUTER FORMULATION FOR THE THREE-DIMENSIONAL TRUSS

There are very few changes that need to be made to the two-dimensional truss program to convert it to a three-dimensional program. Additional input data includes the z-coordinate of each node and a third restraint code, say KZRES(I). The number of degrees of freedom is three times the number of nodes. Of course, a statement incorporating KZRES(I) will have to be added when calculating the number of constraints for determining the size of the reduced structural stiffness matrix. The elements of the elemental stiffness matrices in global coordinates will have to be rewritten in terms of the direction cosines of each member. Loop indices will expand to those of the two-dimensional frame since the number of degrees of freedom is identical. For determination of the member forces, only P_4 will need to be computed. Having completed this much of the text, you should have little difficulty in quickly completing a three-dimensional truss program.

5.6 SUMMARY

In this chapter we developed the necessary transformation matrix for the three-dimensional truss element and expressed the elemental stiffness matrix transformed to the global coordinates in terms of the direction cosines between the member axis and the global coordinate axes. We solved two simple truss problems, including one involving thermal strains, and we discussed the changes that will have to be made to the two-dimensional truss program in order to expand its capability to three dimensions.

PROBLEMS

5.1 A guyed transmission tower is shown in Figure P5-1. If the area of the cross section of the tower is 4 in^2 and the guy wires 1 in^2, find the member forces for the loading shown. $E = 29 \times 10^3$ ksi.

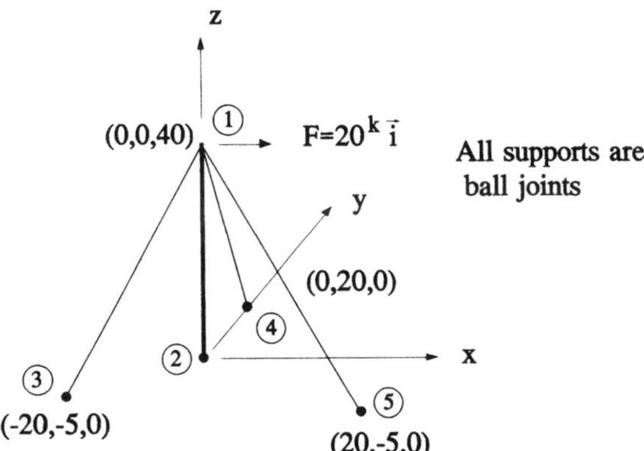

Figure P5-1

5.2 For the structure shown in Figure P5-2, find all bar forces. All areas are 2 in^2 and $E = 29 \times 10^3$ ksi.

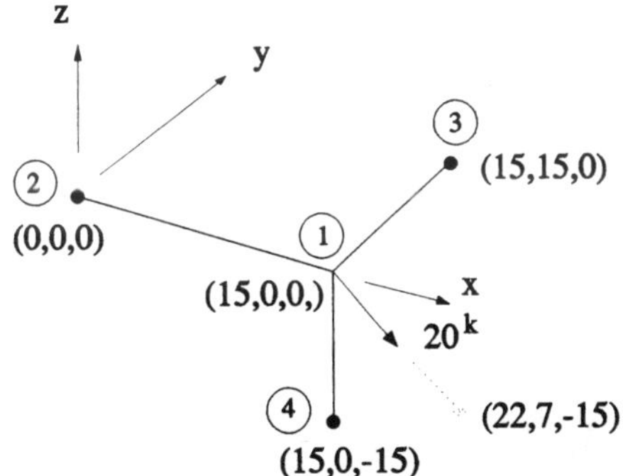

Figure P5-2

5.3 A balcony extension is to be framed as shown in Figure P5-3. Find all bar forces. The areas of the bars are shown in () and $E = 29 \times 10^3$ ksi.

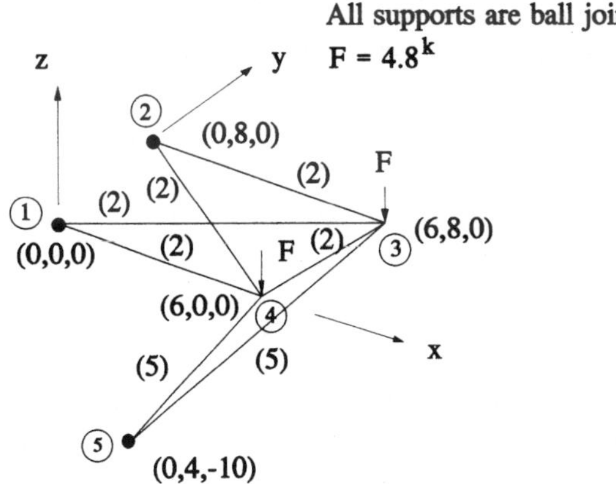

Figure P5-3

5.4 Referring to example problem E5-2, assume that member 1-3 is the only member subjected to the temperature change. Solve for all bar forces.

5.5 Modify your two-dimensional truss program to create a three-dimensional truss program. Use the suggestions in section 5.5 as a guide.

Solve the following problems for nodal displacements and member forces using your three-dimensional truss program as an aid.

5.6 The aircraft engine mount shown in Figure P5-6 is subjected to a torque that results in $F = 800$ lb. If all areas are 1 in^2 and $E = 10 \times 10^3$ ksi, find the bar forces.

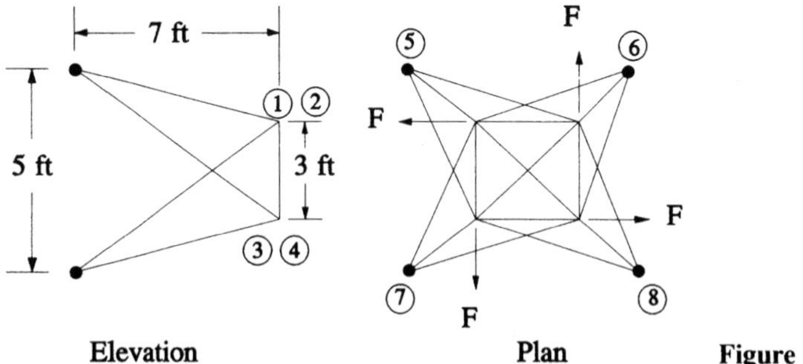

Figure P5-6

5.7 The areas of all bars in the square cross-sectional truss shown in Figure P5-7 are 1 in^2. Find all bar forces. $E = 29 \times 10^3$ ksi.

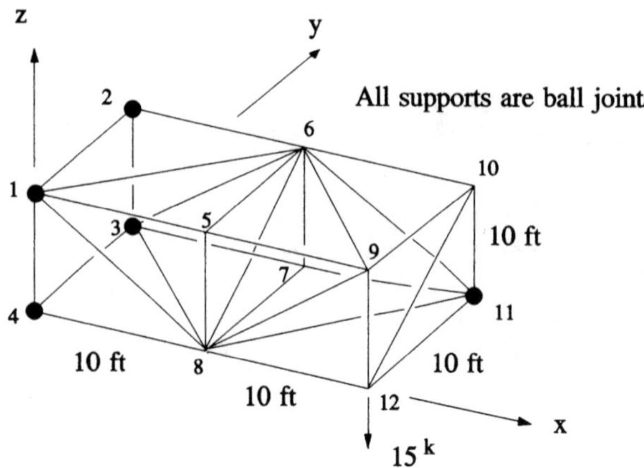

Figure P5-7

CHAPTER 6

ANALYSIS OF THREE-DIMENSIONAL FRAMES

6.1 INTRODUCTION

The three-dimensional frame element is the most complex element that we can treat using strength of materials and standard structural theory. As was the case for the two-dimensional frame element, we assume that the members are connected together with moment-resisting joints. We will also assume that the members are bi-symmetric. That is, the cross section is symmetric about both the local y and z axes. The three-dimensional frame element is an essential element for analysis and design of three-dimensional moment-resisting frames.

To describe the displacements of a node of a three-dimensional frame member, we need to specify three rotations (two bending rotations and a torsional rotation) and three translations (x, y, and z). Thus the frame member will have six degrees of freedom per node, resulting in an elemental stiffness matrix that is of order 12×12. Extending the force and displacement numbering method used for the elements previously considered, we shall number the forces and displacements from 1 to 6 at the left end of the member and from 7 to 12 at the right end.

6.2 DEVELOPMENT OF THE ELEMENTAL STIFFNESS MATRIX

Since we have previously developed the two-dimensional frame element and the grid element, we can easily combine the elemental stiffnesses of these elements to obtain the stiffness matrix for the three-dimensional frame element. If we assume that the elemental coordinate system is oriented as shown in Figure 6-1, note that we will have bending about the y-axis in addition to bending about the z-axis. Twisting about the x-axis and translations in all three coordinate directions will also take place. The positive directions of these displacements and the corresponding forces are shown in Figure 6-1.

Note that subscripts 1, 2, 3, 7, 8, and 9 refer to linear forces and translations, and that subscripts 4, 5, 6, 10, 11, and 12 refer to moments and rotations.

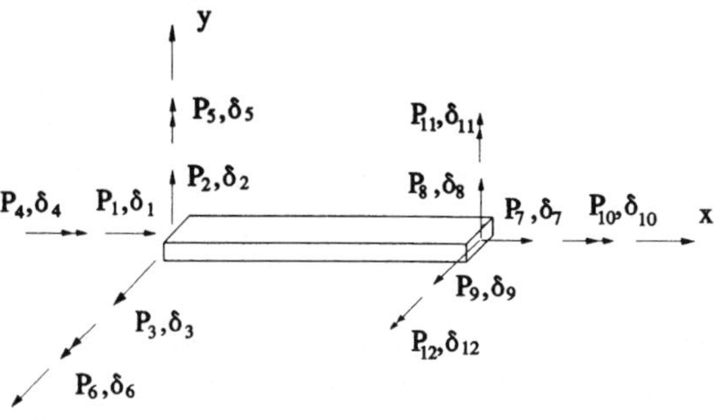

Figure 6-1 Positive directions of elemental forces and displacements.

As with all the previous elements, the elemental stiffness matrix is developed by introducing a unit displacement corresponding to one degree of freedom at a time. The axial and torsional effects are uncoupled from the bending effects as they were for the two-dimensional frame and grid elements. Noting that we have bending about the y and z axes and twisting about the x-axis, we can use the results obtained previously to construct the three-dimensional frame elemental stiffness matrix. Note that we must be careful to subscript the moments of inertia about the y and z axes properly. The resulting elemental stiffness matrix is shown below.

$$\begin{bmatrix}
EA/L & 0 & 0 & 0 & 0 & 0 & -EA/L & 0 & 0 & 0 & 0 & 0 \\
- & 12EI_z/L^3 & 0 & 0 & 0 & 6EI_z/L^2 & 0 & -12EI_z/L^3 & 0 & 0 & 0 & 6EI_z/L^2 \\
- & - & 12EI_y/L^3 & 0 & -6EI_y/L^2 & 0 & 0 & 0 & -12EI_y/L^3 & 0 & -6EI_y/L^2 & 0 \\
- & - & - & GJ_x/L & 0 & 0 & 0 & 0 & 0 & -GJ_x/L & 0 & 0 \\
- & - & - & - & 4EI_y/L & 0 & 0 & 0 & 6EI_y/L^2 & 0 & 2EI_y/L & 0 \\
- & - & - & - & - & 4EI_z/L & 0 & -6EI_z/L^2 & 0 & 0 & 0 & 2EI_z/L \\
- & - & - & - & - & - & EA/L & 0 & 0 & 0 & 0 & 0 \\
- & - & sym. & - & - & - & - & 12EI_z/L^3 & 0 & 0 & 0 & -6EI_z/L^2 \\
- & - & - & - & - & - & - & - & 12EI_y/L^3 & 0 & 6EI_y/L^2 & 0 \\
- & - & - & - & - & - & - & - & - & GJ_x/L & 0 & 0 \\
- & - & - & - & - & - & - & - & - & - & 4EI_y/L & 0 \\
- & - & - & - & - & - & - & - & - & - & - & 4EI_z/L
\end{bmatrix} \tag{6.1}$$

You should note that the elemental stiffness matrices for all elements previously considered can be obtained by retaining the appropriate rows and columns of equation (6.1).

Referring to Figure 6-1, if we want to generate the elemental stiffness matrix for a two-dimensional beam, we would retain only rows and columns corresponding to coordinates 2, 6, 8, and 12. For the grid element, rows and columns 4, 6, 2, 10, 12, and 8 would be retained. Care must be taken, however, to account for different local coordinate labeling and direction.

6.3 TRANSFORMATION OF COORDINATES

As in previous cases, we need to transform the elemental stiffness matrix with respect to the elemental coordinate system to the global coordinate system before combining the stiffnesses to create the structural stiffness matrix. The basic form of this transformation remains the same and is shown in equation (6.2).

$$[K]_{global} = [\beta]^T [k]_{elemental} [\beta] \tag{6.2}$$

In Chapter 5 we recognized that the elements of the $[\beta]$ matrix were the direction cosines of the force and displacement vectors. For the three-dimensional frame element the $[\beta]$ matrix expands to that shown in equation (6.3).

$$[\beta] = \begin{bmatrix} [L] & 0 & 0 & 0 \\ 0 & [L] & 0 & 0 \\ 0 & 0 & [L] & 0 \\ 0 & 0 & 0 & [L] \end{bmatrix} \tag{6.3}$$

where each $[L]$ matrix is the 3×3 matrix of direction cosines.

We now proceed to determine expressions for the direction cosines.

The cosines required are generally calculated from the coordinates of three points. Two of these points are the nodes of the member, and the third point lies in the local x-y plane and is often called the K node. These points are shown in Figure 6-2. As we have seen, the direction cosines are the cosines of the angles between the local coordinate axes and the global axes. For example, l_{23} is the cosine of the angle between the local y-axis and the global z-axis. If we designate the nodes at the ends of the member as A and B we have, as in the case of the three-dimensional truss,

$$l_{11} = (x_B - x_A)/AB, \quad l_{12} = (y_B - y_A)/AB, \quad \text{and } l_{13} = (z_B - z_A)/AB$$

where AB is the length of the member.

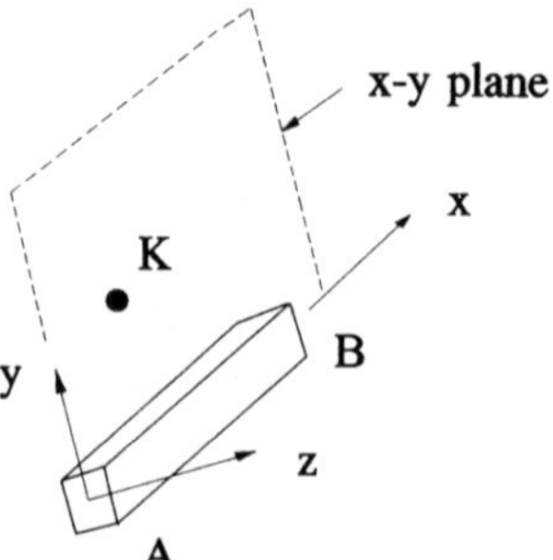

Figure 6-2 Location of K node.

The direction cosines of the local z-axis can be found by realizing that any vector $\vec{\mathbf{Z}}$ parallel to the local z-axis must be perpendicular to the plane formed by two vectors in the local x-y plane. For these two vectors we can use a vector $\vec{\mathbf{X}}$ from point A to point B and a vector $\vec{\mathbf{K}}$ from point A to point K as shown in Figure 6-3. The cross-product between these vectors will yield a vector perpendicular to the x-y plane. Dividing this vector by its length results in a unit vector the components of which are the required direction cosines between the local z-axis and the global axes.

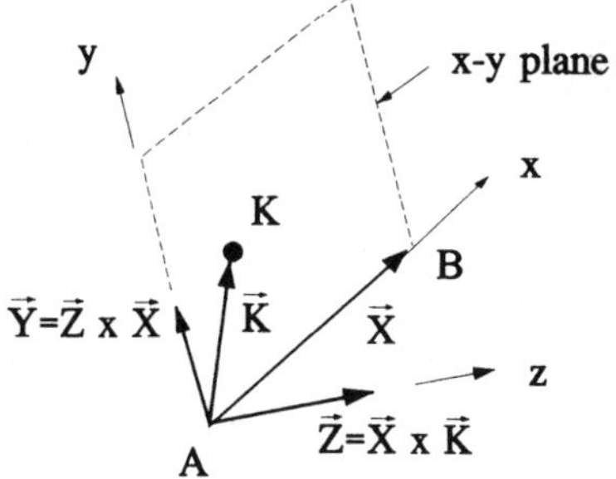

Figure 6-3 Vector products.

The vector $\vec{\mathbf{Z}} = \vec{\mathbf{X}} \times \vec{\mathbf{K}}$. The components of the vectors $\vec{\mathbf{X}}$ and $\vec{\mathbf{K}}$ are given by equation (6.4):

$$\begin{aligned} \vec{X} &= (x_B - x_A)\vec{i} + (y_B - y_A)\vec{j} + (z_B - z_A)\vec{k} \\ \vec{K} &= (x_K - x_A)\vec{i} + (y_K - y_A)\vec{j} + (z_K - z_A)\vec{k} \end{aligned} \tag{6.4}$$

Forming the cross-product yields the following components:

$$\begin{aligned} Z_x &= (y_B - y_A)(z_K - z_A) - (z_B - z_A)(y_K - y_A) \\ Z_y &= (z_B - z_A)(x_K - x_A) - (x_B - x_A)(z_K - z_A) \\ Z_z &= (x_B - x_A)(y_K - y_A) - (y_B - y_A)(x_K - x_A) \end{aligned} \tag{6.5}$$

Thus, $l_{31} = Z_x/Z$, $l_{32} = Z_y/Z$, and $l_{33} = Z_z/Z$ where $Z = \sqrt{Z_x^2 + Z_y^2 + Z_z^2}$

Similarly, the cross-product between a vector in the z-axis direction and one in the x-axis direction will yield a vector in the y-axis direction. The z-axis vector is $l_{31}\vec{\mathbf{i}} + l_{32}\vec{\mathbf{j}} + l_{33}\vec{\mathbf{k}}$ and the x-axis vector is $l_{11}\vec{\mathbf{i}} + l_{12}\vec{\mathbf{j}} + l_{13}\vec{\mathbf{k}}$. Forming the cross-product $\vec{\mathbf{Z}} \times \vec{\mathbf{X}}$ yields the following components:

$$\begin{aligned} Y_x &= l_{13}l_{32} - l_{12}l_{33} \\ Y_y &= l_{11}l_{33} - l_{13}l_{31} \\ Y_z &= l_{12}l_{31} - l_{11}l_{32} \end{aligned} \tag{6.6}$$

Thus, $l_{21} = Y_x/Y$, $l_{22} = Y_y/Y$, and $l_{23} = Y_z/Y$ where $Y = \sqrt{Y_x^2 + Y_y^2 + Y_z^2}$.

Note that the positive direction of the local y-axis will always be perpendicular to the x-axis in the direction of the K node.

Having determined the elements of the $[L]$ matrix, we can transform the elemental stiffnesses to the global coordinate system by using equation (6.2). We then combine them into the structural stiffness matrix as before. We next reduce the structural stiffness matrix by imposing boundary conditions and solve for the unknown displacements. Finally, the member forces are calculated by using equation (6.7).

$$\{P\} = [k]_{element}\{\delta\} = [k]_{element}[\beta]\{u\} \tag{6.7}$$

6.4 EXAMPLE OF A THREE-DIMENSIONAL FRAME PROBLEM

Consider the rigid frame shown in Figure E6-1.

All members are W 8 × 24 steel wide flange sections with the following properties:

$E = 29 \times 10^6$ psi, $G = 11.15 \times 10^6$ psi, $I_z = 82.8$ in^4, $I_y = 18.3$ in^4, $J = 0.35$ in^4, $A = 7.08$ in^2.

The nodal coordinates in inches are:

Node 1: 180,0,180

All members are 15 ft. long.

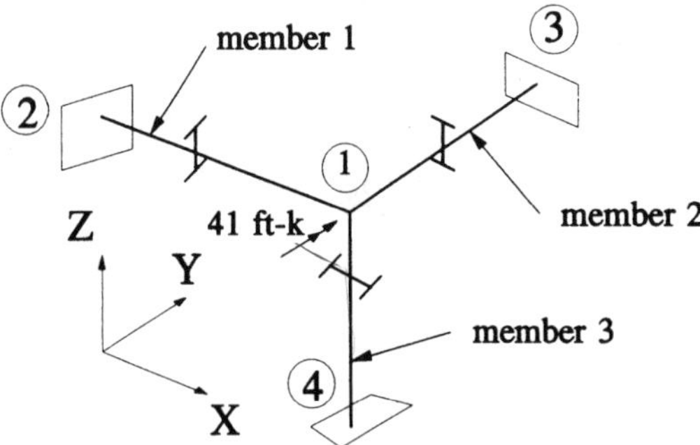

Figure E6-1 Example frame.

Node 2: 0,0,180
Node 3: 180,180,180
Node 4: 180,0,0

Node number 1 will be considered the left node of each member.
We now must select a K node for each member.
The K node coordinates selected for each member are:

Member 1: 0,0,0
Member 2: 180,0,0
Member 3: 0,0,0

Note that the K nodes are located in the local x-y plane for each member. The local axes and the K vectors for each member are shown in Figures E6-2a, E6-2b, and E6-2c.

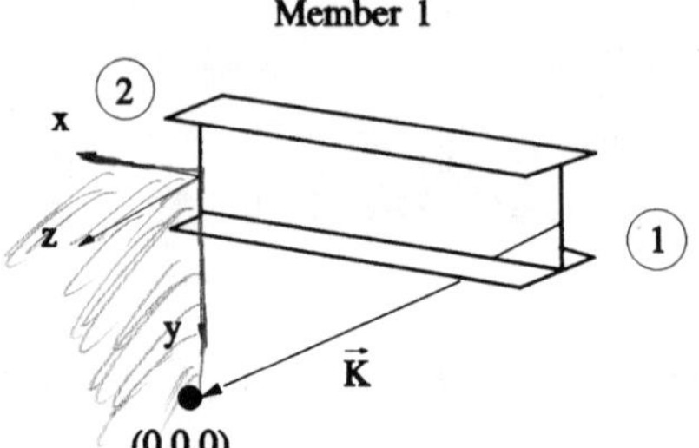

Figure E6-2a K vector for member 1.

To further explain the K node position, consider member number 1 shown in Figure E6-2a.

Since node 1 is the left end of this member, the local x-axis lies along the axis of the member in the direction from node 1 toward node 2. The position of the K node is

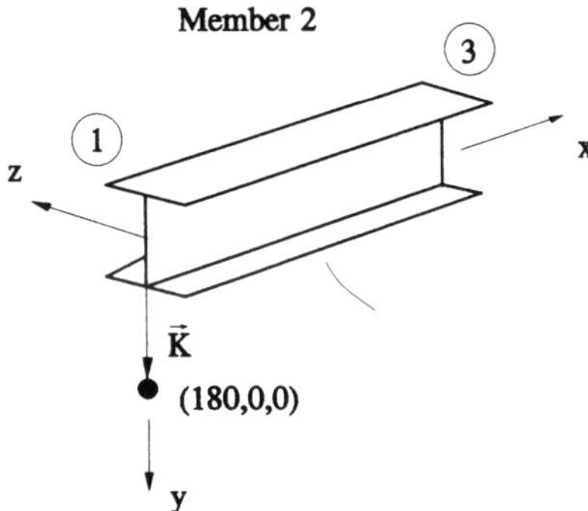

Figure E6-2b K vector for member 2.

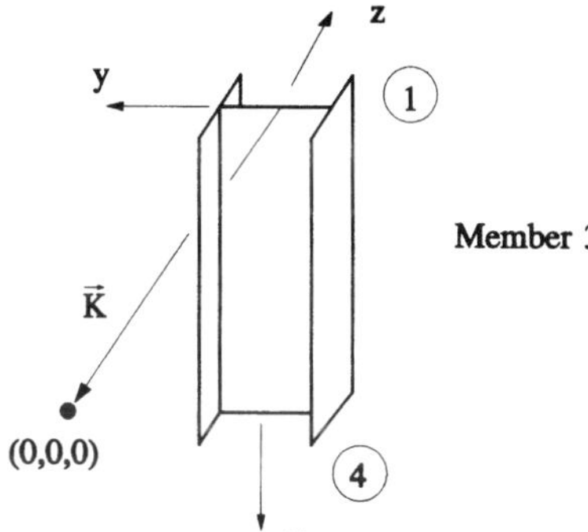

Figure E6-2c K vector for member 3.

selected to lie in the plane of the web of the beam element and is directed from node 1 toward the selected point. In this case, the point at coordinates $(0, 0, 0)$ is chosen as the K node. The y-axis is always perpendicular to the x-axis, and its positive direction is toward the K node. This defines the y-axis in Figure E6-2a. The z-axis positive direction is determined by applying the right-hand rule to the x and y axes.

We next calculate the direction cosines for each member.

The left end of each member is point A for the member. For this example, node 1 is the left end of each member. Thus $x_A = x_1$, $y_A = y_1$, and $z_A = z_1$.

Member 1:

$$
\begin{array}{lll}
x_A = 180 & x_B = 0 & x_K = 0 \\
y_A = 0 & y_B = 0 & y_K = 0 \\
z_A = 180 & z_B = 180 & z_K = 0
\end{array}
$$

$$\mathbf{AB} = \sqrt{(0 - 180)^2 + 0 + 0} = 180$$

$$l_{11} = (0 - 180)/180 = -1 \quad l_{12} = l_{13} = 0$$

From equation (6.5),

$$Z_x = (0)(0 - 180) - (0)(0) = 0$$

$$Z_y = (0)(180 - 0) - (0 - 180)(0 - 180) = -(180)^2$$

$$Z_z = (0 - 180)(0) - (0)(0 - 180) = 0$$

$$Z = (180)^2$$

Thus,

$$l_{31} = 0, \quad l_{32} = -1, \quad l_{33} = 0$$

From equation (6.6),

$$Y_x = 0 - 0 = 0$$

$$Y_y = -1(0) - 0 = 0$$

$$Y_z = 0 - (-1)(-1) = -1$$

Thus,

$$l_{21} = l_{22} = 0, \quad l_{23} = -1$$

The $[L]$ matrix for member 1 becomes

$$[L] = \begin{bmatrix} -1 & 0 & 0 \\ 0 & 0 & -1 \\ 0 & -1 & 0 \end{bmatrix} \tag{6.8}$$

Member 2:

$$x_A = 180 \quad x_B = 180 \quad x_K = 180$$

$$y_A = 0 \quad y_B = 180 \quad y_K = 0$$

$$z_A = 180 \quad z_B = 180 \quad z_K = 0$$

$$AB = 180$$

$$l_{11} = (180 - 180)/180 = 0, l_{12} = (180 - 0)/180 = 1, l_{13} = (180 - 180)/180 = 0$$

From equation (6.5),

$$Z_x = (180)(-180) - 0 = -(180)^2$$

$$Z_y = 0 - 0 = 0$$

$$Z_z = 0 - 0 = 0$$

Thus,

$$l_{31} = -(180)^2/(180)^2 = -1, \quad l_{32} = l_{33} = 0$$

From equation (6.6),

$$Y_x = 0 - 0 = 0$$
$$Y_y = 0 - 0 = 0$$
$$Y_z = 1(-1) - 0 = -1$$

Thus,

$$l_{21} = l_{22} = 0, \quad l_{23} = -1$$

Therefore, for member 2,

$$[L] = \begin{bmatrix} 0 & 1 & 0 \\ 0 & 0 & -1 \\ -1 & 0 & 0 \end{bmatrix} \tag{6.9}$$

Member 3:

$$x_A = 180 \quad x_B = 180 \quad x_K = 0$$
$$y_A = 0 \quad y_B = 0 \quad y_K = 0$$
$$z_A = 180 \quad z_B = 0 \quad z_K = 0$$
$$AB = 180$$
$$l_{11} = 0, l_{12} = 0, l_{13} = -180/180 = -1$$

From equation (6.5),

$$Z_x = 0 - 0 = 0$$
$$Z_y = -180(-180) - 0 = (180)^2$$
$$Z_z = 0 - 0 = 0$$

Thus,

$$l_{31} = l_{33} = 0, \quad l_{32} = 1$$

From equation (6.6),

$$Y_x = -1 - 0 = -1$$
$$Y_y = 0 - 0 = 0$$
$$Y_z = 0 - 0 = 0$$

Thus,

$$l_{21} = -1, \quad l_{22} = l_{23} = 0$$

Therefore, for member 3,

$$[L] = \begin{bmatrix} 0 & 0 & -1 \\ -1 & 0 & 0 \\ 0 & 1 & 0 \end{bmatrix} \tag{6.10}$$

Since node number 1 is the only node with non-zero displacements, and since it also represents the left node of each member, only the top left 6 × 6 of the transformed elemental stiffness matrix for each member is required. Using equation (6.2) for each member we find (for the top left 6 × 6) for each member:

Member 1:

$$\begin{bmatrix} 1{,}140{,}667 & 0 & 0 & 0 & 0 & 0 \\ - & 1{,}092 & 0 & 0 & 0 & -98{,}278 \\ - & - & 4{,}941 & 0 & 444{,}667 & 0 \\ - & sym. & - & 21{,}681 & 0 & 0 \\ - & - & - & - & 53{,}360{,}000 & 0 \\ - & - & - & - & - & 11{,}793{,}333 \end{bmatrix} \tag{6.11}$$

Member 2:

$$\begin{bmatrix} 1{,}092 & 0 & 0 & 0 & 0 & -98{,}278 \\ - & 1{,}140{,}667 & 0 & 0 & 0 & 0 \\ - & - & 4{,}941 & 444{,}667 & 0 & 0 \\ - & sym. & - & 53{,}360{,}000 & 0 & 0 \\ - & - & - & - & 21{,}681 & 0 \\ - & - & - & - & - & 11{,}793{,}333 \end{bmatrix} \tag{6.12}$$

Member 3:

$$\begin{bmatrix} 4{,}941 & 0 & 0 & 0 & -444{,}667 & 0 \\ - & 1{,}092 & 0 & 98{,}278 & 0 & 0 \\ - & - & 1{,}140{,}667 & 0 & 0 & 0 \\ - & sym. & - & 11{,}793{,}333 & 0 & 0 \\ - & - & - & - & 53{,}360{,}000 & 0 \\ - & - & - & - & - & 21{,}681 \end{bmatrix} \tag{6.13}$$

Combining the above equations, the overall structural stiffness equation becomes

$$\begin{Bmatrix} 0 \\ 0 \\ 0 \\ 0 \\ 492{,}000 \; in\text{-}\# \\ 0 \end{Bmatrix} = \begin{bmatrix} 1{,}146{,}699 & 0 & 0 & 0 & -444{,}667 & -98{,}278 \\ - & 1{,}142{,}851 & 0 & 98{,}278 & 0 & -98{,}278 \\ - & - & 1{,}150{,}548 & 444{,}667 & 444{,}667 & 0 \\ - & sym. & - & 65{,}175{,}010 & 0 & 0 \\ - & - & - & - & 106{,}741{,}700 & 0 \\ - & - & - & - & - & 23{,}608{,}350 \end{bmatrix} \begin{Bmatrix} u_1 \\ u_2 \\ u_3 \\ u_4 \\ u_5 \\ u_6 \end{Bmatrix} \tag{6.14}$$

Solving equation (6.14) for the displacements we find

$$\begin{Bmatrix} u_1 \\ u_2 \\ u_3 \\ u_4 \\ u_5 \\ u_6 \end{Bmatrix} = \begin{Bmatrix} .001794 \; in \\ 0 \; in \\ -.001792 \; in \\ .000012 \; rad \\ .004624 \; rad \\ .000007 \; rad \end{Bmatrix} \tag{6.15}$$

Using equation (6.7) for each member, the member forces become

Member 1:

$$\begin{Bmatrix} P_1 \\ P_2 \\ P_3 \\ P_4 \\ P_5 \\ P_6 \\ P_7 \\ P_8 \\ P_9 \\ P_{10} \\ P_{11} \\ P_{12} \end{Bmatrix} = \begin{Bmatrix} -2046 \; \# \\ -2047 \; \# \\ .73\# \\ -.27 \; in\text{-}\# \\ -88 \; in\text{-}\# \\ -245{,}950 \; in\text{-}\# \\ 2046 \; \# \\ 2047 \; \# \\ -.73\# \\ .27 \; in\text{-}\# \\ -44 \; in\text{-}\# \\ -122{,}577 \; in\text{-}\# \end{Bmatrix} \tag{6.16}$$

Member 2:

$$\begin{Bmatrix} P_1 \\ P_2 \\ P_3 \\ P_4 \\ P_5 \\ P_6 \\ P_7 \\ P_8 \\ P_9 \\ P_{10} \\ P_{11} \\ P_{12} \end{Bmatrix} = \begin{Bmatrix} -.47 \; \# \\ 3.42 \; \# \\ -1.23 \; \# \\ 100.3 \; in\text{-}\# \\ 88.3 \; in\text{-}\# \\ 144.4 \; in\text{-}\# \\ .47 \; \# \\ -3.42 \; \# \\ 1.23 \; \# \\ -100.3 \; in\text{-}\# \\ 132.27 \; in\text{-}\# \\ 470.6 \; in\text{-}\# \end{Bmatrix} \tag{6.17}$$

Member 3:

$$\begin{Bmatrix} P_1 \\ P_2 \\ P_3 \\ P_4 \\ P_5 \\ P_6 \\ P_7 \\ P_8 \\ P_9 \\ P_{10} \\ P_{11} \\ P_{12} \end{Bmatrix} = \begin{Bmatrix} 2{,}044\ \# \\ 2{,}047\ \# \\ 1.2\ \# \\ -.16\ in\text{-}\# \\ -144.2\ in\text{-}\# \\ 245{,}949\ in\text{-}\# \\ -2{,}044\ \# \\ -2{,}047\ \# \\ -1.2\ \# \\ .16\ in\text{-}\# \\ -72.05\ in\text{-}\# \\ 122{,}576\ in\text{-}\# \end{Bmatrix} \tag{6.18}$$

The forces (in units of pounds and inch-pounds) at the left end of each member are shown in Figure E6-3. Note that the torsional moments are very small in comparison to the bending moments. This occurs since the torsional stiffness of these open cross sections is small compared to the bending stiffness.

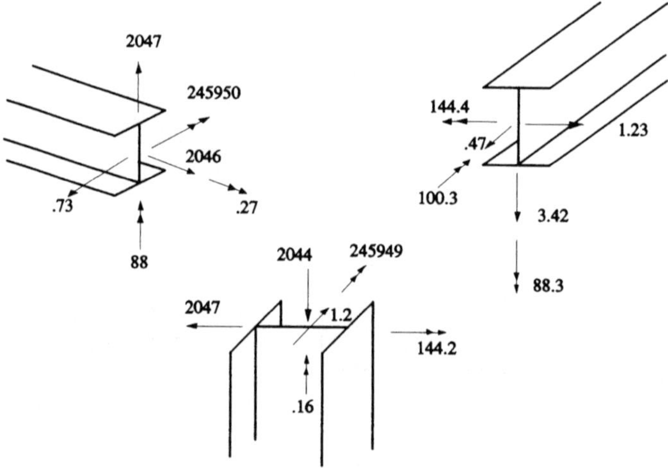

Figure E6-3 Member forces.

6.5 COMPUTER FORMULATION

Since the matrices we must deal with are quite large (12 × 12), it will be more convenient to perform the necessary matrix multiplications numerically rather than derive the general case as we did for all previous elements. Of course, this will take considerably more computer time and require more storage for the intermediate matrices generated when transforming the elemental stiffnesses to global coordinates. The notation for displacements follows the same convention as for the previous elements. That is, the displacement subscripts will be designated as six times the node number and six times the node number minus 1, 2, 3, 4, and 5. There will be six restraint codes for each node, and the order of the reduced structural stiffness matrix will be six times the number of nodes minus the number of restrained displacements.

Matrix multiplication is defined in Appendix A as

$$c_{ij} = \sum_{k=1}^{N} a_{ik} b_{kj} \tag{6.19}$$

A code fragment to perform this operation is shown below.

```
FOR I=1 TO N     (N=12 for the three-dimensional frame element)
FOR J=1 TO N
C(I,J)=0
FOR K=1 TO N
C(I,J)=A(I,K)*B(K,J)+C(I,J)
NEXT K: NEXT J: NEXT I
```

The $[\beta]$ matrix can be transposed using a code fragment similar to that shown below.

```
FOR I=1 TO N
FOR J=I TO N
TEMP=B(I,J)
B(I,J)=B(J,I)
B(J,I)=TEMP
NEXT J: NEXT I
```

After the reduced stiffness matrix has been inverted and the unknown displacements found, the member forces are found as usual by using $\{P\} = [k]_{\text{element}}[\beta]\{u\}$. Remember that the first six displacements are those of the left end of the member and the last six are those of the right end of the member. These will be represented by 6*ML(I)-5, 6*ML(I)-4, ... etc., and 6*MR(I)-5, 6*MR(I)-4 ... etc.

6.6 SUMMARY

In this chapter we investigated the use of a three-dimensional frame element for solving structural problems. We have seen that the elements of the $[\beta]$ transformation matrix can be found using the coordinates of the ends of the member and a third K node that lies in the local x-y plane of the element. An example frame problem was solved and computer formulation of a three-dimensional frame program was discussed. Of course, non-nodal loads are addressed in the same manner as for the previous elements discussed; that is, by using the concept of equivalent nodal loads. In addition, support settlements are dealt with by using the same techniques presented earlier in the text.

In the chapters that follow we shall address additional topics in the stiffness formulation of structural problems and alternate ways of deriving the elemental stiffness matrices, deriving the overall structural stiffness equation, and treating the topic of non-nodal loads.

PROBLEMS

$E = 29 \times 10^6$ psi and $G = 11.2 \times 10^6$ psi for all problems.

6.1 All members of the frame shown in Figure P6-1 are square structural tube with the following properties: $I_y = I_z = 27$ in^4, $J_x = 46.8$ in^4, $A = 8.36$ in^2. A moment vector with components of -36 ft-k, 0, and 48 ft-k acts at node 1. Find all displacements and member forces. The coordinates shown are in feet.

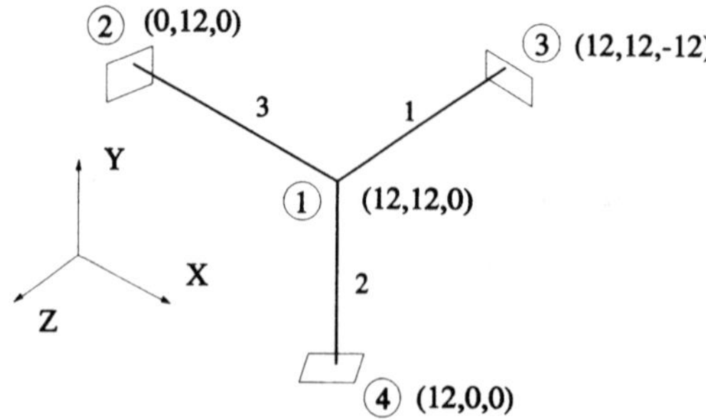

Figure P6-1

6.2 All members of the frame shown in Figure P6-2 have the same properties as those of problem P6.1. Find all displacements and member forces. The coordinates are given in feet.

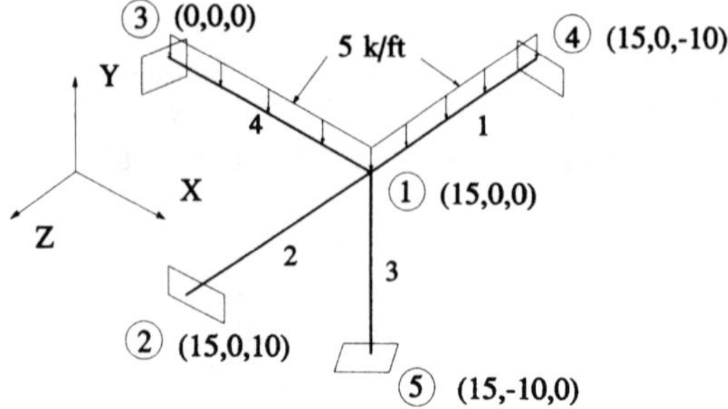

Figure P6-2

6.3 Using the suggestions given in section 6.5, write a three-dimensional frame computer program. It should have the capability of solving problems with 25 members and 20 joints. *Hint:* Modify your two-dimensional frame program.

Use your computer program from problem 6.3 as the basis for solution of the following problems.

6.4 All frame members in Figure P6-4 are W 12×22 with the following properties: $I_y = 4.66$ in^4, $I_z = 156$ in^4, $J_x = 0.29$ in^4, $A = 6.48$ in^2. Find all displacements and member forces.

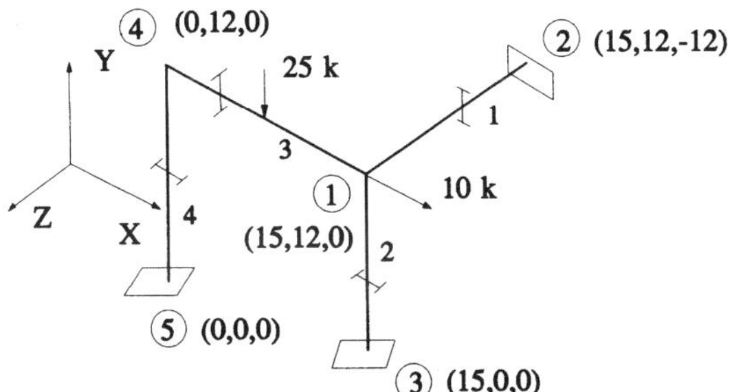

Figure P6-4

6.5 All members in Figure P6-5 are W 12 × 22 (see problem 6.4 for properties). Find all displacements and member forces.

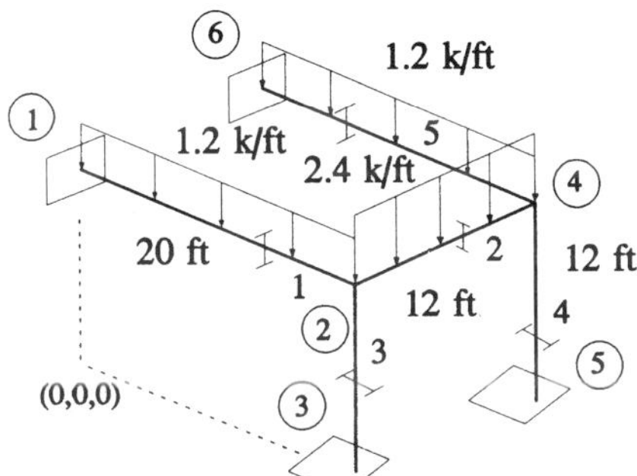

Figure P6-5

6.6 All members of the frame shown in Figure P6-6 are square tube with $A = 27.4$ in^2, $I_y = I_z = 580$ in^4, $J_x = 943$ in^4. Find all member forces.

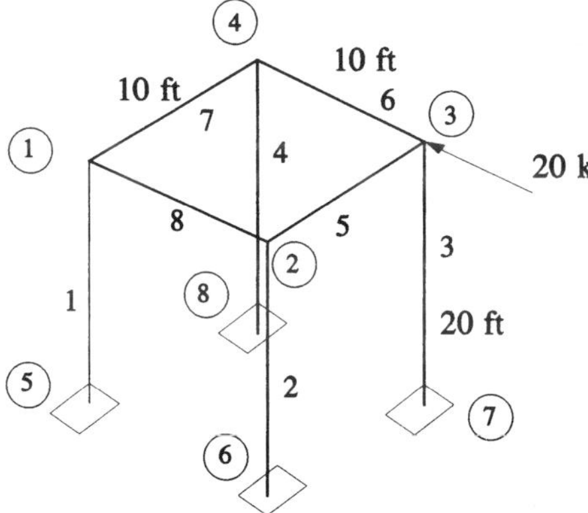

Figure P6-6

6.7 All members of the frame shown in Figure P6-7 are square tube with $A = 6.58$ in^2, $I_y = I_z = 22.8$ in^4, $J_x = 38.2$ in^4. For the two inclined members, the z-axis of the tube cross section lies in the plane formed by these members. Find all displacements and member forces.

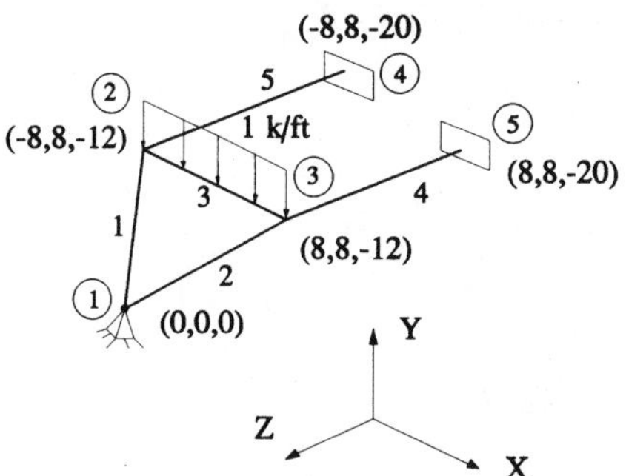

Figure P6-7

6.8 The members of the frame shown in Figure P6-8 are made of square tube with $A = 3.11$ in^2, $I_y = I_z = 3.58$ in^4, $J_x = 3.32$ in^4. Each of the four applied moments shown has a magnitude of 1.6 ft-k. Find all member forces.

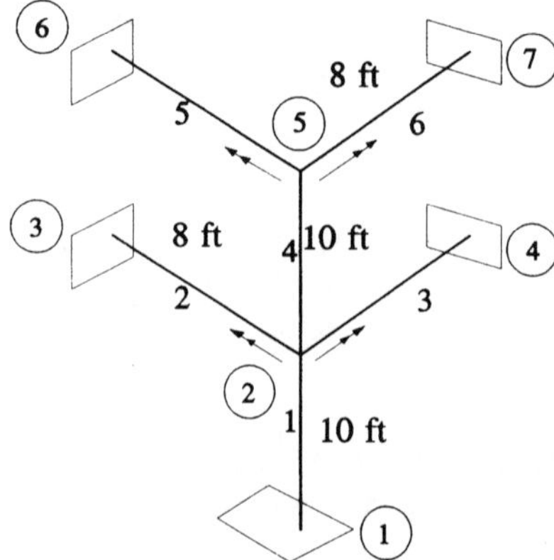

Figure P6-8

CHAPTER 7

ADDITIONAL TOPICS IN THE STIFFNESS METHOD

7.1 DISCUSSION OF BANDWIDTH

We have seen that the global force-displacement equation for the linear-elastic structures considered in this text contains a stiffness matrix that is symmetric and has positive terms on the main diagonal. The storage requirements for the global stiffness matrix could certainly be reduced by taking advantage of this symmetry. That is, we need to store only the main diagonal and upper or lower portion above or below the main diagonal. Computational efficiency would also be increased by using an inversion routine that took account of this symmetry.

We can, however, decrease storage requirements even more by recognizing that most non-zero elements are clustered around the main diagonal, forming a diagonal "band" of matrix elements. The width of this band (bandwidth) is dependent on the nodal numbering scheme chosen. To illustrate this, consider the following frame with two different nodal numbering methods.

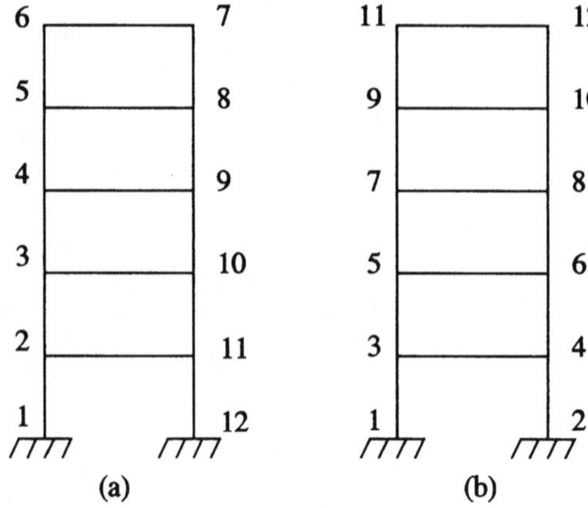

Figure 7-1 Two frame node numbering schemes.

Since we are dealing with a two-dimensional frame, each node has three degrees of freedom, two translations and a rotation. Also, each member stiffness matrix is a 6×6 matrix with 3×3 submatrices associated with the left and right nodal displacements. That is, the top left and right 3×3 submatrices are associated with the left node forces and left and right nodal displacements, respectively. Similarly, the bottom left and right 3×3 matrices are associated with the right node forces and the left and right nodal displacements. For example, if we denote these 3×3 submatrices with an X, the member that connects nodes 6 and 7 in Figure 7-1a will contribute to the overall structural stiffness matrix in locations (6,6), (7,7), (6,7), and (7,6). Filling in our 12×12 matrix of 3×3 submatrices X, we find for Figure 7-1a:

$$[K] = \begin{bmatrix} X & X & - & - & - & - & - & - & - & - & - & - \\ X & X & X & - & - & - & - & - & - & - & X & - \\ - & X & X & X & - & - & - & - & - & X & - & - \\ - & - & X & X & X & - & - & - & X & - & - & - \\ - & - & - & X & X & X & - & X & - & - & - & - \\ - & - & - & - & X & X & X & - & - & - & - & - \\ - & - & - & - & - & X & X & X & - & - & - & - \\ - & - & - & - & X & - & X & X & X & - & - & - \\ - & - & - & X & - & - & - & X & X & X & - & - \\ - & - & X & - & - & - & - & - & X & X & X & - \\ - & X & - & - & - & - & - & - & - & X & X & X \\ - & - & - & - & - & - & - & - & - & - & X & X \end{bmatrix} \tag{7.1}$$

For Figure 7-1b we have:

$$[K] = \begin{bmatrix} X & - & X & - & - & - & - & - & - & - & - & - \\ - & X & - & X & - & - & - & - & - & - & - & - \\ X & - & X & X & X & - & - & - & - & - & - & - \\ - & X & X & X & - & X & - & - & - & - & - & - \\ - & - & X & - & X & X & X & - & - & - & - & - \\ - & - & - & X & X & X & - & X & - & - & - & - \\ - & - & - & - & X & - & X & X & X & - & - & - \\ - & - & - & - & - & X & X & X & - & X & - & - \\ - & - & - & - & - & - & X & - & X & X & X & - \\ - & - & - & - & - & - & - & X & X & X & - & X \\ - & - & - & - & - & - & - & - & X & - & X & X \\ - & - & - & - & - & - & - & - & - & X & X & X \end{bmatrix} \tag{7.2}$$

Since we have a symmetric matrix, we need to store elements on the main diagonal and those non-zero terms to the right (or left) of the main diagonal. This total width is called the "half-bandwidth" or sometimes the "semi-bandwidth." Note that for Figure 7-1a, the half-bandwidth is $10 \times 3 = 30$ (remember that each X represents a 3×3 matrix), while for the numbering scheme of Figure 7-1b the half-bandwidth is $3 \times 3 = 9$. This means that our storage requirements would be either a 36×30 matrix or a 36×9 matrix. Clearly, we can reduce our storage needs by more than a factor of three simply by carefully numbering the nodes.

We want to minimize the maximum difference in node numbers between connected nodes. This generally means numbering the nodes across the short direction of the structure as in Figure 7-1b.

The general expression for half-bandwidth is given by

$$\text{bw} = dof(maxdiff + 1) \tag{7.3}$$

where

bw = half-bandwidth

dof = number of degrees of freedom per node

maxdiff = maximum difference in node numbers between connected nodes.

The half-bandwidths for the elements considered in this text are listed below:

One-dimensional bar	bw = 1(maxdiff) + 1
Two-dimensional truss	bw = 2(maxdiff) + 2
Two-dimensional frame	bw = 3(maxdiff) + 3
Grid	bw = 3(maxdiff) + 3
Three-dimensional truss	bw = 3(maxdiff) + 3
Three-dimensional frame	bw = 6(maxdiff) + 6

There are matrix inversion routines that operate on symmetric matrices stored in half-bandwidth form. Not only are storage requirements reduced, but because there are many fewer numbers to manipulate, the inversion process occurs much more quickly.

It is a simple task to build the structural stiffness matrix in half-bandwidth form. The following code segment does this for the two-dimensional frame. We assume that the elemental stiffness matrices have been transformed to global coordinates.

```
FOR I=1 to NM        [loop on number of members]
IJ(1)=3*ML(I)-2:IJ(2)=3*ML(I)-1:IJ(3)=3*ML(I)
IJ(4)=3*MR(I)-2:IJ(5)=3*MR(I)-1:IJ(6)=3*MR(I)
FOR IR=1 TO 6       [row index]
FOR IC=1 TO 6       [col index]
IF(IJ(IC)<IJ(IR)) THEN 200
CC=IJ(IC)-IJ(IR)+1
SK(IJ(IR),CC)=SK(IJ(IR),CC)+EKT(IR,IC)
200 NEXT IC:NEXT IR: NEXT I
```

Since the column order is not maintained when we store the matrix in half-bandwidth form, we must use a different technique to account for displacement boundary conditions. A simple way to deal with the boundary conditions is to multiply the first element in the row associated with the displacement boundary condition, which is the main diagonal element, by a large number and set the force associated with the displacement equal to the displacement multiplied by the resulting product of the main diagonal element and the large number. This is exactly the same technique as presented in section 1.7 of Chapter 1. This procedure also works with zero displacement boundary conditions.

7.2 COMBINING DIFFERENT ELEMENTS TO MODEL A STRUCTURE

Often it is necessary to use more than one type of element in order to analyze a structure. An example is shown in Figure 7-2 where a rigid frame is braced by truss members.

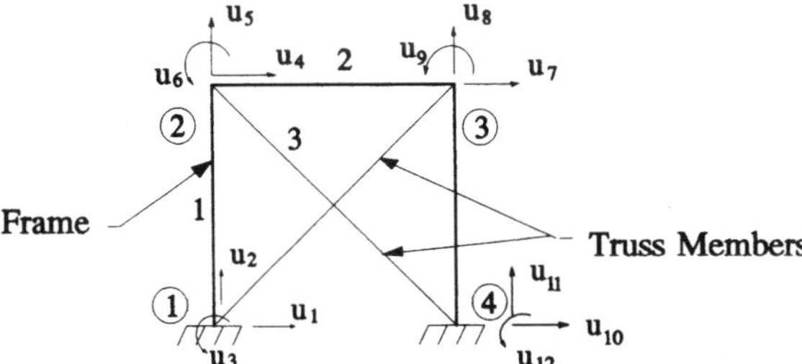

Figure 7-2 Structure with frame and truss members.

Considering joint 2, we notice that there will be contributions to the global stiffness matrix from frame members 1 and 2 and the truss element, member 3. Since we have three degrees of freedom at each node for the frame members (two translations and one rotation) and only two for the truss element (two translations), we must take care to ensure that the elements of the truss stiffness matrix are placed in the correct locations in the global structural stiffness matrix. This is illustrated in the following example.

Example 7.1

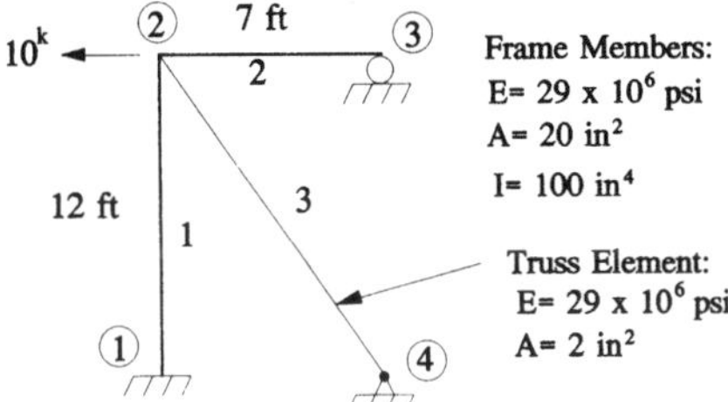

Figure 7-3 Example 7.1.

The stiffness matrices of members 1, 2, and 3 shown in Figure 7-3, with respect to the global coordinate system, are given below:

Member 1:

$$
\begin{array}{cccccc}
① & ② & ③ & ④ & ⑤ & ⑥
\end{array}
$$

$$
\begin{bmatrix}
11.65 & 0 & -839.12 & -11.65 & 0 & -839.12 \\
- & 4027.78 & 0 & 0 & -4027.78 & 0 \\
- & - & 80555.55 & 839.12 & 0 & 4027.78 \\
- & - & - & 11.65 & 0 & 839.12 \\
- & sym. & - & - & 4027.78 & 0 \\
- & - & - & - & - & 80555.55
\end{bmatrix}
\begin{array}{c} ① \\ ② \\ ③ \\ ④ \\ ⑤ \\ ⑥ \end{array}
\tag{7.4}
$$

Member 2:

$$\begin{array}{c} \begin{matrix} \text{④} & \text{⑤} & \text{⑥} & \text{⑦} & \text{⑧} & \text{⑨} \end{matrix} \\ \begin{bmatrix} 6904.76 & 0 & 0 & -6904.76 & 0 & 0 \\ - & 58.71 & 2465.99 & 0 & -58.71 & 2465.99 \\ - & - & 138095.2 & 0 & -2465.99 & 69047.62 \\ - & - & - & 6904.76 & 0 & 0 \\ - & sym. & - & - & 58.71 & -2465.99 \\ - & - & - & - & - & 138095.2 \end{bmatrix} \begin{matrix} \text{④} \\ \text{⑤} \\ \text{⑥} \\ \text{⑦} \\ \text{⑧} \\ \text{⑨} \end{matrix} \end{array} \tag{7.5}$$

Member 3:

$$\begin{array}{c} \begin{matrix} \text{④} & \text{⑤} & \text{⑩} & \text{⑪} \end{matrix} \\ \begin{bmatrix} 88.33 & -151.42 & -88.33 & 151.42 \\ - & 259.58 & 151.42 & -259.58 \\ - & - & 88.33 & -151.42 \\ - & sym. & - & 259.58 \end{bmatrix} \begin{matrix} \text{④} \\ \text{⑤} \\ \text{⑩} \\ \text{⑪} \end{matrix} \end{array} \tag{7.6}$$

Note that the global displacement numbers corresponding to the elements of the above matrices are shown above the columns and to the right of the rows.

Also note that the truss elemental stiffness will contribute to the rows and columns corresponding to 3*node number -2 and 3*node number -1; that is, the translational components of the nodal displacements.

Proceeding to formulate the reduced structural stiffness matrix by adding appropriate elemental stiffness terms and removing rows and columns associated with zero displacement boundary conditions, we obtain:

$$\begin{bmatrix} 7004.75 & -151.42 & 839.12 & -6904.76 & 0 \\ - & 4346.07 & 2465.99 & 0 & 2465.99 \\ - & - & 218650.75 & 0 & 69047.62 \\ - & sym. & - & 6904.76 & 0 \\ - & - & - & - & 138095.2 \end{bmatrix} \tag{7.7}$$

Inverting and solving for displacements with $F_4 = -10^k$ as the only applied force, we find

$$\begin{Bmatrix} u_4 \\ u_5 \\ u_6 \\ u_7 \\ u_9 \end{Bmatrix} = \begin{Bmatrix} -.110592\ in \\ -.004045\ in \\ .000531\ rad \\ -.110592\ rad \\ -.000193\ rad \end{Bmatrix} \tag{7.8}$$

For the truss element $\{\delta\} = [\beta]\{u\}$

$$\delta_1 = .50387(-.110592) - .86378(-.004045) = -.05223 \text{ in}$$

$$\delta_3 = 0$$

$$\{P\} = [k]_e\{\delta\}$$

$$P_3 = 347.91(-\delta_1) = 347.91(.05223) = 18.17^k$$

One way to automate the above process in a computer program would be to accumulate elements of the overall global stiffness matrix in separate subroutines for each type of element used to model the structure. Of course, the loop indices used would have to reflect the appropriate global displacement number. In this example, in the truss member subroutine, IJ(1)=3*ML(I)-2, IJ(2)=3*ML(I)-1, IJ(3)=3*MR(I)-2, and IJ(4)=3*MR(I)-1.

An alternative to the procedure just described is to treat all members as frame members but to use a very small moment of inertia for the truss member in comparison to those of the frame members. Although this technique results in an approximate solution, it generally is a very good one. For the above example, using a moment of inertia for the truss member of 1 in^4 and treating it as a frame member yields the following solution for truss member forces:

$$\begin{Bmatrix} P_1 \\ P_2 \\ P_3 \\ P_4 \\ P_5 \\ P_6 \end{Bmatrix} = \begin{Bmatrix} -18.17\ k \\ 0\ k \\ -.03\ ft\text{-}k \\ 18.17\ k \\ 0\ k \\ 0\ ft\text{-}k \end{Bmatrix} \tag{7.9}$$

Notice that the axial force is the same as that of the previous solution and that the shear forces and bending moments are effectively zero, as they should be.

7.3 ELASTIC SUPPORTS

The simplest procedure for treating linearly elastic supports, whether axial or torsional, is to introduce an additional element at the support that has the required spring constant or constants.

Example 7.2

Consider the beam shown in Figure 7-4. The member force-displacement relationships are given by equations (7.10) and (7.11). The spring relationship is given by equation (7.12).

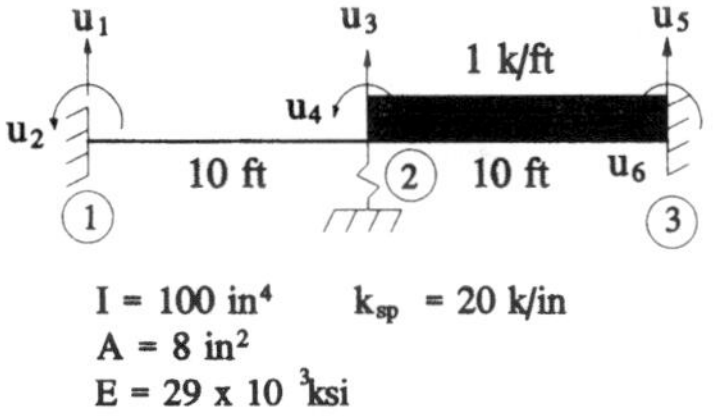

u1
u3
u5
1 k/ft
u2
u4
10 ft
2
10 ft
u6
1
3
I = 100 in⁴
k_sp = 20 k/in
A = 8 in²
E = 29 x 10 ³ksi

Figure 7-4 Beam with an elastic support.

Member 1:

$$\begin{Bmatrix} F_1 \\ F_2 \\ F_3 \\ F_4 \end{Bmatrix} = \begin{bmatrix} 20.14 & 1208.33 & -20.14 & 1208.33 \\ - & 96666.66 & -1208.33 & 48333.33 \\ sym. & - & 20.14 & -1208.33 \\ - & - & - & 96666.66 \end{bmatrix} \begin{Bmatrix} u_1 \\ u_2 \\ u_3 \\ u_4 \end{Bmatrix} \tag{7.10}$$

Member 2:

$$\begin{Bmatrix} F_3 \\ F_4 \\ F_5 \\ F_6 \end{Bmatrix} = \begin{bmatrix} 20.14 & 1208.33 & -20.14 & 1208.33 \\ - & 96666.66 & -1208.33 & 48333.33 \\ sym. & - & 20.14 & -1208.33 \\ - & - & - & 96666.66 \end{bmatrix} \begin{Bmatrix} u_3 \\ u_4 \\ u_5 \\ u_6 \end{Bmatrix} \tag{7.11}$$

Spring:

$$F_3 = k_{spring} u_3 \tag{7.12}$$

Combining to generate the reduced stiffness matrix we find

$$\begin{Bmatrix} F_3 \\ F_4 \end{Bmatrix} = \begin{bmatrix} k_{33}^{(1)} + k_{11}^{(2)} + k_{sp} & k_{34}^{(1)} + k_{12}^{(2)} \\ k_{43}^{(1)} + k_{21}^{(2)} & k_{44}^{(1)} + k_{22}^{(2)} \end{bmatrix} \begin{Bmatrix} u_3 \\ u_4 \end{Bmatrix} \tag{7.13}$$

Numerically we have

$$\begin{Bmatrix} F_3 \\ F_4 \end{Bmatrix} = \begin{bmatrix} 60.28 & 0 \\ 0 & 193333.3 \end{bmatrix} \begin{Bmatrix} u_3 \\ u_4 \end{Bmatrix} \tag{7.14}$$

The equivalent nodal forces at node 2, F_3 and F_4, are -5^k and $-100''^{-k}$. Solving for the displacements we find

$$\begin{Bmatrix} u_3 \\ u_4 \end{Bmatrix} = \begin{bmatrix} .01659 & 0 \\ 0 & 5.1724 \times 10^{-6} \end{bmatrix} \begin{Bmatrix} -5 \\ -100 \end{Bmatrix} = \begin{Bmatrix} -.082949 \\ -.000517 \end{Bmatrix} \tag{7.15}$$

An alternate method of solution treats the structure as a frame and defines the spring as an additional member where its properties are adjusted to approximate a one-dimensional member.

In the previous example we want $EA/L = 20$ k/in. Using $E = 29 \times 10^3$ ksi and $L = 120$ in, the area A must equal .08276 in^2 in order to meet the axial stiffness requirement. By using a very small moment of inertia I of .01 in^4, the bending stiffness is negligible in comparison to those of the beam elements. After solution, the moments and shears in the member representing the spring are found to be zero to two decimal places. The displacements are identical to those of equation (7.15).

Naturally, these techniques can also be used to model torsional springs or a combination of axial and torsional springs. For example, if the beam shown in Figure 7-4 had a torsional spring at node 2, the $F_4 = k_{torsion} u_4$ would replace equation (7.12). The torsional spring stiffness would be added to the element of the overall stiffness matrix that related F_4 to u_4.

If the frame approximation is used to represent a torsional spring, two vertical members can be added as shown in Figure 7-5. The torsional stiffness of these two members in this case is given by $k_{torsion} = 8EI/L$. Two members are used instead of one so as to have the net shear in these members add to zero at the support. The areas selected for these members would be large in comparison to those of the actual frame members to approximate zero vertical displacement at this node.

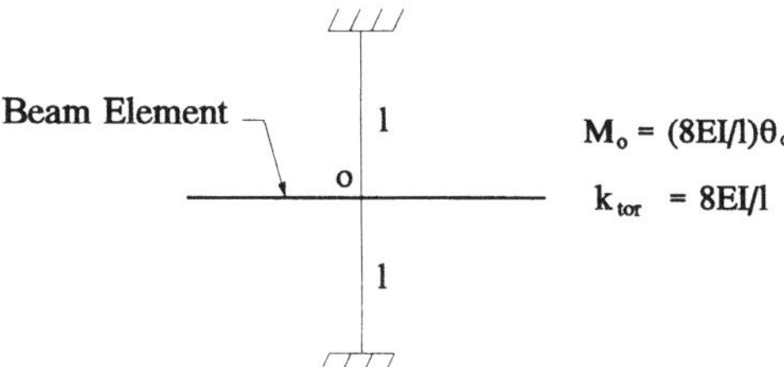

Figure 7-5 Torsional spring approximation.

7.4 INCLINED SUPPORTS

Up to this point in the text we have dealt with support conditions that restrict or specify motion in the global coordinate directions only. However, situations arise similar to the one shown in Figure 7-6.

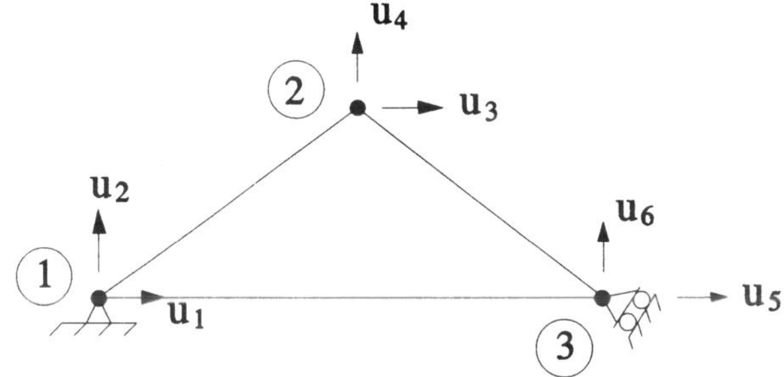

Figure 7-6 Inclined support.

The reaction at node 3 must be perpendicular to the support, and the displacement in that direction is zero. We want to modify our force-displacement relationships to use forces and displacements parallel and perpendicular to the support; that is, in the directions of a local coordinate system.

Consider Figure 7-7, which shows the global and local displacements at node 3.

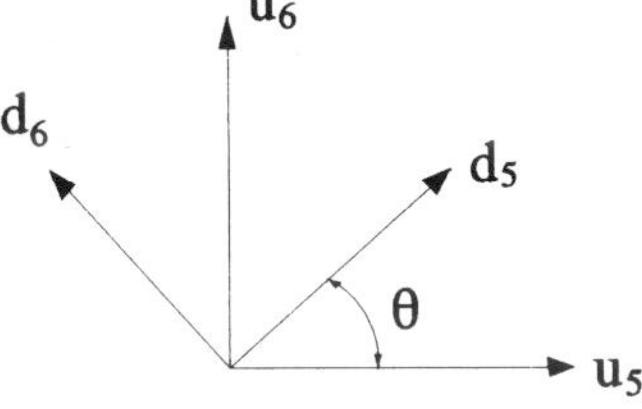

Figure 7-7 Global and local displacements of node 3.

Just as $\{\delta\} = [\beta]\{u\}$, we can write

$$\begin{Bmatrix} d_5 \\ d_6 \end{Bmatrix} = \begin{bmatrix} \cos\theta & \sin\theta \\ -\sin\theta & \cos\theta \end{bmatrix} \begin{Bmatrix} u_5 \\ u_6 \end{Bmatrix} \tag{7.16}$$

Thus,

$$\begin{Bmatrix} u_5 \\ u_6 \end{Bmatrix} = \begin{bmatrix} \cos\theta & -\sin\theta \\ \sin\theta & \cos\theta \end{bmatrix} \begin{Bmatrix} d_5 \\ d_6 \end{Bmatrix} = [\beta]^T \begin{Bmatrix} d_5 \\ d_6 \end{Bmatrix} \tag{7.17}$$

Since both u_1 and u_2 are zero for the structure, we can write

$$\begin{Bmatrix} u_3 \\ u_4 \\ u_5 \\ u_6 \end{Bmatrix} = \begin{bmatrix} 1 & 0 & 0 & 0 \\ 0 & 1 & 0 & 0 \\ 0 & 0 & \cos\theta & -\sin\theta \\ 0 & 0 & \sin\theta & \cos\theta \end{bmatrix} \begin{Bmatrix} u_3 \\ u_4 \\ d_5 \\ d_6 \end{Bmatrix} = [\beta'] \begin{Bmatrix} u_3 \\ u_4 \\ d_5 \\ d_6 \end{Bmatrix} \tag{7.18}$$

Now,

$$\begin{Bmatrix} F_3 \\ F_4 \\ F_5 \\ F_6 \end{Bmatrix} = [K] \begin{Bmatrix} u_3 \\ u_4 \\ u_5 \\ u_6 \end{Bmatrix} = [K][\beta'] \begin{Bmatrix} u_3 \\ u_4 \\ d_5 \\ d_6 \end{Bmatrix} \tag{7.19}$$

The forces are also vectors and we can write

$$\begin{Bmatrix} F_3 \\ F_4 \\ F_5 \\ F_6 \end{Bmatrix} = \begin{bmatrix} 1 & 0 & 0 & 0 \\ 0 & 1 & 0 & 0 \\ 0 & 0 & \cos\theta & -\sin\theta \\ 0 & 0 & \sin\theta & \cos\theta \end{bmatrix} \begin{Bmatrix} F_3 \\ F_4 \\ R_5 \\ R_6 \end{Bmatrix} = [\beta'] \begin{Bmatrix} F_3 \\ F_4 \\ R_5 \\ R_6 \end{Bmatrix} \tag{7.20}$$

where R_5 and R_6 are forces corresponding to the displacements d_5 and d_6.

From equation (7.20),

$$\begin{Bmatrix} F_3 \\ F_4 \\ R_5 \\ R_6 \end{Bmatrix} = [\beta']^T \begin{Bmatrix} F_3 \\ F_4 \\ F_5 \\ F_6 \end{Bmatrix} \tag{7.21}$$

Using equation (7.19) in equation (7.21) we have

$$\begin{Bmatrix} F_3 \\ F_4 \\ R_5 \\ R_6 \end{Bmatrix} = [\beta']^T [K][\beta'] \begin{Bmatrix} u_3 \\ u_4 \\ d_5 \\ d_6 \end{Bmatrix} \tag{7.22}$$

At this point we can enforce the condition that $d_6 = 0$, reduce the equation, and solve for the unknown displacements.

Example 7.3

Consider the truss shown in Figure 7-8.

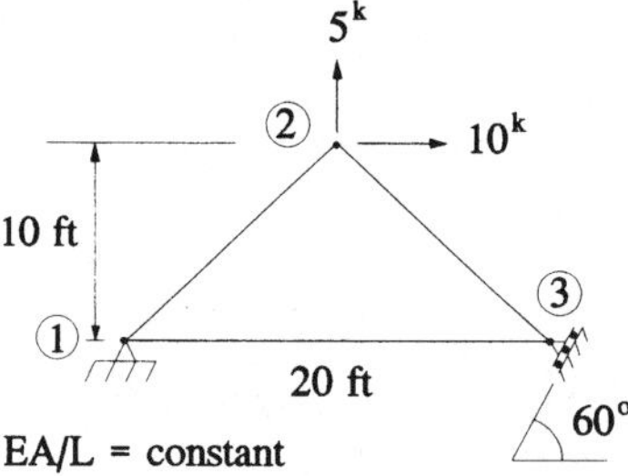

Figure 7-8 Example 7.3.

The structural stiffness matrix, reduced by using the zero displacement conditions at node 1, but not yet accounting for the support at node 3, is

$$[K] = EA/l \begin{bmatrix} 1 & 0 & -.5 & .5 \\ 0 & 1 & .5 & -.5 \\ -.5 & .5 & 1.5 & -.5 \\ .5 & -.5 & -.5 & .5 \end{bmatrix} \tag{7.23}$$

For the support at node 3, $\theta = 60^0$ and $[\beta']$ from equation (7.18) becomes

$$[\beta'] = \begin{bmatrix} 1 & 0 & 0 & 0 \\ 0 & 1 & 0 & 0 \\ 0 & 0 & .5 & -.866 \\ 0 & 0 & .866 & .5 \end{bmatrix} \tag{7.24}$$

Using equation (7.24) in equation (7.22) yields

$$\begin{Bmatrix} F_3 \\ F_4 \\ R_5 \\ R_6 \end{Bmatrix} = EA/l \begin{bmatrix} 1 & 0 & .183 & .683 \\ 0 & 1 & -.183 & -.683 \\ .183 & -.183 & .317 & -.183 \\ .683 & -.683 & -.183 & 1.683 \end{bmatrix} \begin{Bmatrix} u_3 \\ u_4 \\ d_5 \\ d_6 \end{Bmatrix} \tag{7.25}$$

We now enforce the condition $d_6 = 0$ by eliminating the fourth row and column. Thus,

$$\begin{Bmatrix} 10 \\ 5 \\ 0 \end{Bmatrix} = EA/l \begin{bmatrix} 1 & 0 & .183 \\ 0 & 1 & -.183 \\ .183 & -.183 & .317 \end{bmatrix} \begin{Bmatrix} u_3 \\ u_4 \\ d_5 \end{Bmatrix} \tag{7.26}$$

Solving equation (7.26) for the displacements yields

$$\begin{Bmatrix} u_3 \\ u_4 \\ d_5 \end{Bmatrix} = l/EA \begin{Bmatrix} 10.67 \\ 4.33 \\ -3.64 \end{Bmatrix} \tag{7.27}$$

The global displacements are found by using equation (7.18).

$$\begin{Bmatrix} u_3 \\ u_4 \\ u_5 \\ u_6 \end{Bmatrix} = l/EA \begin{Bmatrix} 10.67 \\ 4.33 \\ -1.83 \\ -3.17 \end{Bmatrix} \tag{7.28}$$

Another approach to solving problems with inclined supports is to replace the support with an inclined axial force member that is very stiff in comparison to the other members of the structure.

In the previous example, if the axial stiffness EA/L for each member is taken as 241.66 k/in, equation (7.28) yields

$$\begin{Bmatrix} u_3 \\ u_4 \\ u_5 \\ u_6 \end{Bmatrix} = \begin{Bmatrix} .0442 \\ .0179 \\ -.0076 \\ -.0131 \end{Bmatrix} in \tag{7.29}$$

By adding a member in the direction perpendicular to the support, which has an axial stiffness of 29 × 10^3 k/in, the following displacements are found:

$$\begin{Bmatrix} u_3 \\ u_4 \\ u_5 \\ u_6 \end{Bmatrix} = \begin{Bmatrix} .0443 \\ .0178 \\ -.0076 \\ -.0135 \end{Bmatrix} in \qquad (7.30)$$

Equations (7.29) and (7.30) compare very favorably.

7.5 HINGES IN BEAM AND FRAME ELEMENTS

Recall from your basic structural analysis coursework that when a hinge is inserted into a beam or frame structure you have introduced a "release" of a force at the location of the hinge. The bending moment is zero at that point. Remember also that the knowledge of the value of a force at a specific location in a member of a structure yields an "equation of condition." By considering a portion of the structure as a free body by isolating at the hinge, the equation of condition can be used to generate an additional equation that is independent of the overall equilibrium equations. Consider the frame shown in Figure 7-9.

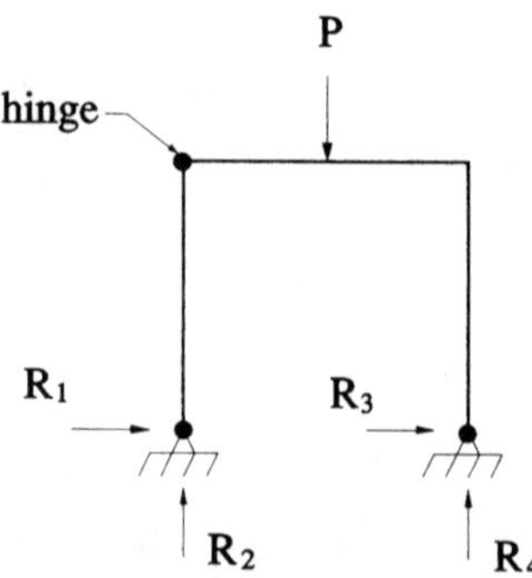

Figure 7-9 Frame with internal hinge.

The frame has four possible reactions and, of course, there are only three overall equilibrium equations available for this two-dimensional structure.

Owing to the presence of the hinge, however, we have an equation of condition available; specifically, the bending moment at the hinge is zero. By disconnecting the structure at the hinge, we obtain the free-body diagrams shown in Figure 7-10.

By using free-body diagrams obtained by isolating the structure at the points where these moments are known, we can write equations that are independent of the overall equilibrium equations of the structure. These independent equations are used in conjunction with the overall equilibrium equations to find the reactions of the structure. Since these bending moments are known forces at specific points in the structure, equations of equilibrium written at these points constitute equations of condition.

The fact that there is no bending moment at the hinge allows us to write an equation of moment equilibrium for either of the free bodies, thus determining that $R_1 = 0$ or

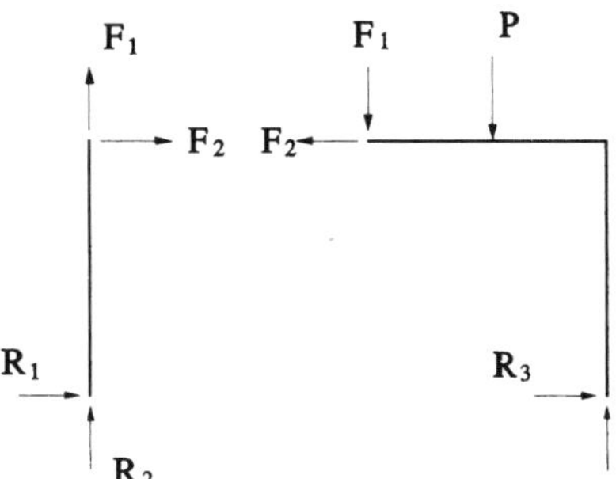

Figure 7-10 Free-body diagrams obtained by disconnecting at hinge.

obtaining a relationship among P, R_3, and R_4. This equation, when used with the overall equilibrium equations, allows us to find all reactions.

We use this technique frequently when solving problems by slope-deflection or moment distribution since we obtain values of the bending moments at the supports using these techniques. Since these bending moments are known forces at specific points of the structure, they can be used to generate equations of condition.

Let us now derive the elemental stiffness matrix for a beam element with a pin at its left end. We will do this directly in this section and by using the technique of static condensation in section 7.6.

The force-displacement relationship for the beam element was derived in section 3.2. It is repeated here for convenience.

$$\begin{Bmatrix} P_1 \\ P_2 \\ P_3 \\ P_4 \end{Bmatrix} = \begin{bmatrix} 12EI/L^3 & 6EI/L^2 & -12EI/L^3 & 6EI/L^2 \\ 6EI/L^2 & 4EI/L & -6EI/L^2 & 2EI/L \\ -12EI/L^3 & -6EI/L^2 & 12EI/L^3 & -6EI/L^2 \\ 6EI/L^2 & 2EI/L & -6EI/L^2 & 4EI/L \end{bmatrix} \begin{Bmatrix} u_1 \\ u_2 \\ u_3 \\ u_4 \end{Bmatrix} \tag{7.31}$$

With a pin at the left node, the bending moment $P_2 = 0$. Expanding the second row of equation (7.31) we have

$$P_2 = \frac{6EI}{l^2}u_1 + \frac{4EI}{l}u_2 - \frac{6EI}{l^2}u_3 + \frac{2EI}{l}u_4 = 0 \tag{7.32}$$

We next solve this equation for u_2 in terms of the other displacements and find

$$u_2 = \frac{-3}{2l}u_1 + \frac{3}{2l}u_3 - \frac{u_4}{2} \tag{7.33}$$

Substituting u_2 from equation (7.33) into equation (7.31) and expanding the first row, we obtain

$$P_1 = \frac{3EI}{l^3}u_1 - \frac{3EI}{l^3}u_3 + \frac{3EI}{l^2}u_4 \tag{7.34}$$

Similarly, for the third and fourth rows we find

$$P_3 = \frac{-3EI}{l^3}u_1 + \frac{3EI}{l^3}u_3 - \frac{3EI}{l^2}u_4 \tag{7.35}$$

$$P_4 = \frac{3EI}{l^2}u_1 - \frac{3EI}{l^2}u_3 + \frac{3EI}{l}u_4 \tag{7.36}$$

Writing the above equations in matrix form yields the following force-displacement relationship for a beam element with a pin at the left node:

$$\begin{Bmatrix} P_1 \\ P_2 \\ P_3 \\ P_4 \end{Bmatrix} = \begin{bmatrix} 3EI/l^3 & 0 & -3EI/l^3 & 3EI/l^2 \\ 0 & 0 & 0 & 0 \\ -3EI/l^3 & 0 & 3EI/l^3 & -3EI/l^2 \\ 3EI/l^2 & 0 & -3EI/l^2 & 3EI/l \end{bmatrix} \begin{Bmatrix} u_1 \\ u_2 \\ u_3 \\ u_4 \end{Bmatrix} \tag{7.37}$$

In equation (7.37), the second row is zero since $P_2 = 0$ and the second column is zero since the effects of u_2 have been incorporated into the other stiffness elements. In other words, the effects of u_2 are implicitly contained in equation (7.37).

Because the axial and bending effects in the frame element are uncoupled, we can modify equation (7.37) by adding the two rows and columns corresponding to the axial effects to obtain the equation for a frame element with a pin at its left end. We have

$$\begin{Bmatrix} P_1 \\ P_2 \\ P_3 \\ P_4 \\ P_5 \\ P_6 \end{Bmatrix} = \begin{bmatrix} EA/l & 0 & 0 & -EA/l & 0 & 0 \\ 0 & 3EI/l^3 & 0 & 0 & -3EI/l^3 & 3EI/l^2 \\ 0 & 0 & 0 & 0 & 0 & 0 \\ -EA/l & 0 & 0 & EA/l & 0 & 0 \\ 0 & -3EI/l^3 & 0 & 0 & 3EI/l^3 & -3EI/l^2 \\ 0 & 3EI/l^2 & 0 & 0 & -3EI/l^2 & 3EI/l \end{bmatrix} \begin{Bmatrix} \delta_1 \\ \delta_2 \\ \delta_3 \\ \delta_4 \\ \delta_5 \\ \delta_6 \end{Bmatrix} \tag{7.38}$$

For a pin at the right end, $P_4 = 0$ for the beam element. Proceeding in the same fashion as for the case with the pin at the left end, we find

$$\begin{Bmatrix} P_1 \\ P_2 \\ P_3 \\ P_4 \end{Bmatrix} = \begin{bmatrix} 3EI/l^3 & 3EI/l^2 & -3EI/l^3 & 0 \\ 3EI/l^2 & 3EI/l & -3EI/l^2 & 0 \\ -3EI/l^3 & -3EI/l^2 & 3EI/l^3 & 0 \\ 0 & 0 & 0 & 0 \end{bmatrix} \begin{Bmatrix} u_1 \\ u_2 \\ u_3 \\ u_4 \end{Bmatrix} \tag{7.39}$$

For the frame element we obtain

$$\begin{Bmatrix} P_1 \\ P_2 \\ P_3 \\ P_4 \\ P_5 \\ P_6 \end{Bmatrix} = \begin{bmatrix} EA/l & 0 & 0 & -EA/l & 0 & 0 \\ 0 & 3EI/l^3 & 3EI/l^2 & 0 & -3EI/l^3 & 0 \\ 0 & 3EI/l^2 & 3EI/l & 0 & -3EI/l^2 & 0 \\ -EA/l & 0 & 0 & EA/l & 0 & 0 \\ 0 & -3EI/l^3 & -3EI/l^2 & 0 & 3EI/l^3 & 0 \\ 0 & 0 & 0 & 0 & 0 & 0 \end{bmatrix} \begin{Bmatrix} \delta_1 \\ \delta_2 \\ \delta_3 \\ \delta_4 \\ \delta_5 \\ \delta_6 \end{Bmatrix} \tag{7.40}$$

Consider the case of a pin at some interior point of a beam element as shown in Figure 7-11.

a b

Figure 7-11 Pin at interior of beam element.

Figure 7-12 shows free-body diagrams considering each portion as a separate beam element.

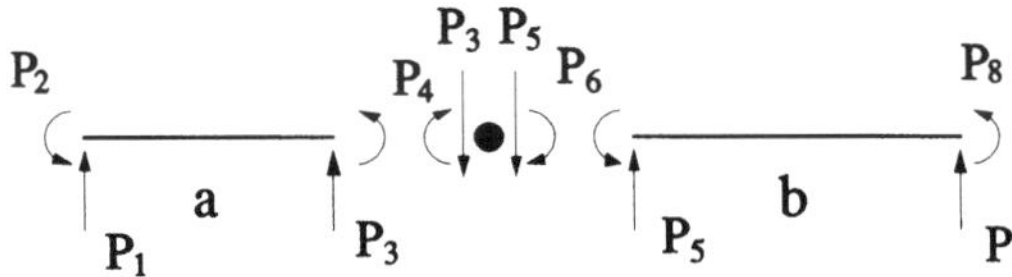

Figure 7-12 Free-body diagrams of left and right segments of beam.

In Figure 7-12, P_4 and P_6 are zero. Vertical equilibrium of the pin requires $P_3 + P_5 = 0$. In addition, $\delta_3 = \delta_5$ for displacement compatibility at the pin.

Treating the left element of Figure 7-12 as a beam with the pin at the right node, and the right element as a beam with the pin at the left, we have

$$\begin{Bmatrix} P_1 \\ P_2 \\ P_3 \\ P_4 \end{Bmatrix} = \begin{bmatrix} 3EI/a^3 & 3EI/a^2 & -3EI/a^3 & 0 \\ 3EI/a^2 & 3EI/a & -3EI/a^2 & 0 \\ -3EI/a^3 & -3EI/a^2 & 3EI/a^3 & 0 \\ 0 & 0 & 0 & 0 \end{bmatrix} \begin{Bmatrix} \delta_1 \\ \delta_2 \\ \delta_3 \\ \delta_4 \end{Bmatrix} \tag{7.41}$$

$$\begin{Bmatrix} P_5 \\ P_6 \\ P_7 \\ P_8 \end{Bmatrix} = \begin{bmatrix} 3EI/b^3 & 0 & -3EI/b^3 & 3EI/b^2 \\ 0 & 0 & 0 & 0 \\ -3EI/b^3 & 0 & 3EI/b^3 & -3EI/b^2 \\ 3EI/b^2 & 0 & -3EI/b^2 & 3EI/b \end{bmatrix} \begin{Bmatrix} \delta_5 \\ \delta_6 \\ \delta_7 \\ \delta_8 \end{Bmatrix} \tag{7.42}$$

By expanding the rows corresponding to P_3 and P_5, using the equations $P_3 + P_5 = 0$ and $\delta_3 = \delta_5$, and solving for δ_3 in terms of the other displacements, we find

$$\delta_3 = \frac{a^3 b^3}{a^3 + b^3} \left[\frac{\delta_1}{a^3} + \frac{\delta_2}{a^2} + \frac{\delta_7}{b^3} - \frac{\delta_8}{b^2} \right] = \delta_5 \tag{7.43}$$

Using equation (7.43) in equations (7.41) and (7.42), we find, after expanding and collecting terms,

$$\begin{Bmatrix} P_1 \\ P_2 \\ P_7 \\ P_8 \end{Bmatrix} = \frac{3EI}{a^3 + b^3} \begin{bmatrix} 1 & a & -1 & b \\ a & a^2 & -a & ab \\ -1 & -a & 1 & -b \\ b & ab & -b & b^2 \end{bmatrix} \begin{Bmatrix} \delta_1 \\ \delta_2 \\ \delta_7 \\ \delta_8 \end{Bmatrix} \tag{7.44}$$

By replacing P_7, P_8, δ_7, and δ_8 in equation (7.44) by P_3, P_4, δ_3, and δ_4, respectively, we have the force-displacement relationship for a beam element with an interior pin or hinge.

Example 7.4

Consider the frame shown in Figure 7-13.

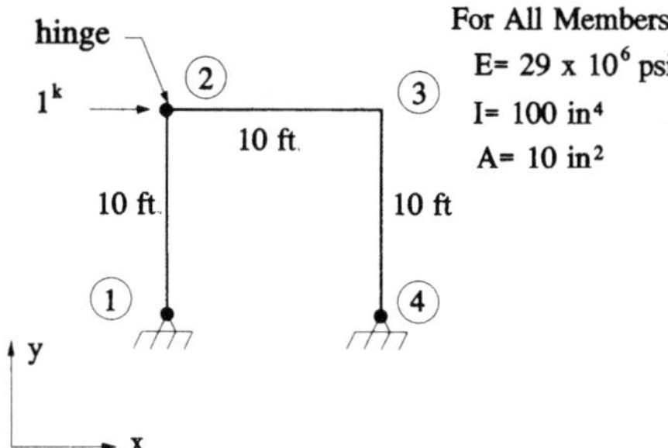

Figure 7-13 Example 7.4.

We shall consider the hinge to be at the left end of member 2-3. Thus, the stiffness matrix in equation (7.38) is used for this member.

After transforming the stiffness of the vertical members to the global coordinate system, combining the three elemental stiffnesses, and accounting for the zero displacement support conditions, our reduced structural stiffness equation becomes

$$\begin{Bmatrix} 0 \\ 1 \\ 0 \\ 0 \\ 0 \\ 0 \\ 0 \\ 0 \end{Bmatrix} = \begin{bmatrix} 4EI/l & 6EI/l^2 & 0 & 2EI/l \\ - & 12EI/l^3 + EA/l & 0 & 6EI/l^2 \\ - & - & 3EI/l^3 + EA/l & 0 \\ - & - & - & 4EI/l \\ - & - & - & - \\ - & sym. & - & - \\ - & - & - & - \\ - & - & - & - \end{bmatrix} \quad \text{continues}$$

$$\begin{matrix} 0 & 0 & 0 & 0 \\ -EA/l & 0 & 0 & 0 \\ 0 & -3EI/l^3 & 3EI/l^2 & 0 \\ 0 & 0 & 0 & 0 \\ 12EI/l^3 + EA/l & 0 & 6EI/l^2 & 6EI/l^2 \\ - & 3EI/l^3 + EA/l & -3EI/l^2 & 0 \\ - & - & 7EI/l & 2EI/l \\ - & - & - & 4EI/l \end{matrix} \Bigg] \begin{Bmatrix} u_3 \\ u_4 \\ u_5 \\ u_6 \\ u_7 \\ u_8 \\ u_9 \\ u_{12} \end{Bmatrix} \tag{7.45}$$

Note that F_4 is the only non-zero applied load.

After evaluating the elements of equation (7.45) numerically, inverting the matrix, and solving for the displacements, we find

$$\begin{Bmatrix} u_3 \\ u_4 \\ u_5 \\ u_6 \\ u_7 \\ u_8 \\ u_9 \\ u_{12} \end{Bmatrix} = \begin{Bmatrix} -.0033 \\ .3985 \\ .00041 \\ -.0033 \\ .3981 \\ -.00041 \\ -.00166 \\ -.00414 \end{Bmatrix} \tag{7.46}$$

Note that u_6 is the rotation at node 2 of member 1-2. To obtain the rotation of member 2-3 at the pin we must use equation (7.33), treating u_1 and u_3 as the vertical displacements at the ends of the member, and u_4 as the rotation at the right end.

Thus, for the rotation at the pin we have

$$\theta = \frac{-3}{2l}(.0004) + \frac{3}{2l}(-.0004) - \left(-\frac{.0017}{2}\right) = .0008 \text{ rad} \tag{7.47}$$

An approximate solution to the problem presented can be obtained by introducing a very short horizontal member at node 2, which has very little bending stiffness. We then use our original frame program for analyzing the structure. For example, introducing a member with a length of 0.1 ft, a cross-sectional area of 10 in^2, and a moment of inertia of 0.001 in^4, as shown in Figure 7-14, yields the following displacements:

$$\begin{Bmatrix} u_3 \\ u_4 \\ u_5 \\ u_6 \\ u_7 \\ u_8 \\ u_9 \\ u_{10} \\ u_{11} \\ u_{12} \\ u_{15} \end{Bmatrix} = \begin{Bmatrix} -.0033 \\ .396 \\ 0 \\ -.0033 \\ .396 \\ .004 \\ .0008 \\ .396 \\ 0 \\ -.0017 \\ -.0041 \end{Bmatrix} \tag{7.48}$$

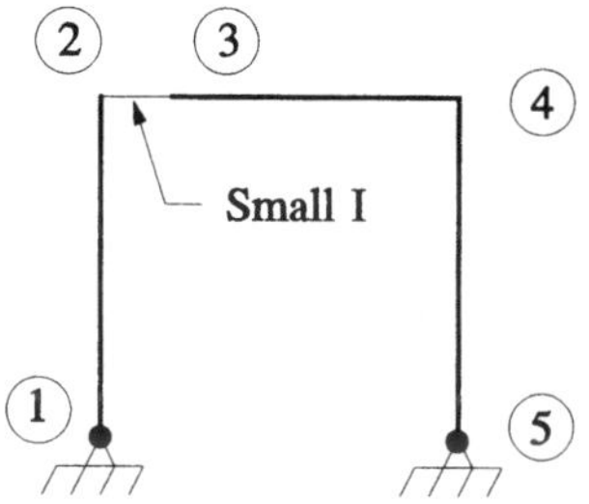

2 3 4 Small I 1 5

Figure 7-14 Approximation for a hinge.

The displacements u_7, u_8, and u_9 are associated with the added node, node 3. Comparison of equations (7.48) and (7.46) shows very good agreement. Displacement u_9 is the rotation at node 3 and is the same value as the previously calculated value of the rotation at the pin (equation [7.47]).

7.6 STATIC CONDENSATION

In the previous section we obtained the force-displacement relationship for beam and frame members with hinges present. This was accomplished by expressing one or more displacements in terms of the others. The solution of a set of simultaneous equations is found using the same procedure. We reduce, or condense, the number of simultaneous equations that need to be solved. The procedure termed *static condensation* does exactly this.

Consider the matrix equation (7.49):

$$\begin{Bmatrix} F_p \\ F_s \end{Bmatrix} = \begin{bmatrix} K_{pp} & K_{ps} \\ K_{sp} & K_{ss} \end{bmatrix} \begin{Bmatrix} u_p \\ u_s \end{Bmatrix} \tag{7.49}$$

Note that we have partitioned this global equation to separate "primary" quantities (designated with a "p" subscript) from "secondary" quantities (designated with an "s" subscript). What we plan to do is eliminate the secondary displacements from this matrix equation. The secondary displacements will be those associated with the equation of condition introduced by the presence of the hinge.

Expanding equation (7.49) we have

$$F_p = K_{pp}u_p + K_{ps}u_s \tag{7.50}$$

$$F_s = K_{sp}u_p + K_{ss}u_s \tag{7.51}$$

We next solve equation (7.51) for u_s giving

$$u_s = K_{ss}^{-1}\,[\,F_s - K_{sp}u_p\,] \tag{7.52}$$

Substituting equation (7.52) into equation (7.50) we find

$$F_p = K_{pp}u_p + K_{ps}K_{ss}^{-1}\,[\,F_s - K_{sp}u_p\,] \tag{7.53}$$

which can be written

$$F_p - K_{ps}K_{ss}^{-1}F_s = (K_{pp} - K_{ps}K_{ss}^{-1}K_{sp})u_p \tag{7.54}$$

or

$$F_{condensed} = K_{condensed}u_p \tag{7.55}$$

After solving equation (7.55) for u_p, equation (7.52) is used to determine u_s.

Note that we are inverting matrices that are smaller than the original structural stiffness matrix. There are, however, a large number of matrix multiplications to perform. The size of the problem to be solved will determine the advantage or disadvantage of using the method.

Example 7.5

Consider a beam with a pin at the right node. For this case, $P_4 = 0$ and we will treat u_4 as a secondary quantity. We therefore partition the original beam force-displacement relationship as shown in equation (7.56).

$$\begin{Bmatrix} P_1 \\ P_2 \\ P_3 \\ \hline P_4 \end{Bmatrix} = \left[\begin{array}{ccc|c} 12EI/L^3 & 6EI/L^2 & -12EI/L^3 & 6EI/L^2 \\ 6EI/L^2 & 4EI/L & -6EI/L^2 & 2EI/L \\ -12EI/L^3 & -6EI/L^2 & 12EI/L^3 & -6EI/L^2 \\ \hline 6EI/L^2 & 2EI/L & -6EI/L^2 & 4EI/L \end{array}\right] \begin{Bmatrix} u_1 \\ u_2 \\ u_3 \\ u_4 \end{Bmatrix} \tag{7.56}$$

Note that F_s in equation (7.49) is P_4 in equation (7.56), which is zero (zero moment at the hinge). The condensed force matrix is simply F_p. Thus, from equations (7.54) and (7.55),

$$F_c = \begin{Bmatrix} P_1 \\ P_2 \\ P_3 \end{Bmatrix} \tag{7.57}$$

Since $K_{ss} = 4EI/L$, a single element, $K_{ss}^{-1} = L/4EI$. Performing the operations indicated on the right-hand side of equation (7.54) we find the condensed stiffness matrix, equation (7.58).

$$K_c = \begin{bmatrix} 3EI/l^3 & 3EI/l^2 & -3EI/l^3 \\ 3EI/l^2 & 3EI/l & -3EI/l^2 \\ -3EI/l^3 & -3EI/l^2 & 3EI/l^3 \end{bmatrix} \tag{7.58}$$

Thus, the condensed force-displacement relationship for a beam element with a hinge at the right node becomes

$$\begin{Bmatrix} P_1 \\ P_2 \\ P_3 \end{Bmatrix} = \begin{bmatrix} 3EI/l^3 & 3EI/l^2 & -3EI/l^3 \\ 3EI/l^2 & 3EI/l & -3EI/l^2 \\ -3EI/l^3 & -3EI/l^2 & 3EI/l^3 \end{bmatrix} \begin{Bmatrix} u_1 \\ u_2 \\ u_3 \end{Bmatrix} \tag{7.59}$$

Equation (7.59) is identical to equation (7.39).

Example 7.6

Solve Example problem 3.3 by using static condensation to treat u_3 as a secondary quantity.

Reordering equation (3.28) to place the terms in the form of equation (7.49), we have

$$\begin{Bmatrix} -1.1232 \times 10^6 \\ 748{,}800 \\ -42{,}120 \end{Bmatrix} = \begin{bmatrix} 580 \times 10^6 & 145 \times 10^6 & 0 \\ 145 \times 10^6 & 290 \times 10^6 & 3.625 \times 10^6 \\ 0 & 3.625 \times 10^6 & 120{,}833 \end{bmatrix} \begin{Bmatrix} u_4 \\ u_6 \\ u_3 \end{Bmatrix} \tag{7.60}$$

From equation (7.54),

$$F_c = \begin{Bmatrix} -1.123 \times 10^6 \\ 748{,}800 \end{Bmatrix} - \begin{Bmatrix} 0 \\ 3.625 \times 10^6 \end{Bmatrix} \frac{(-42{,}120)}{120{,}833} = \begin{Bmatrix} -1.1232 \times 10^6 \\ 2.0124 \times 10^6 \end{Bmatrix} \tag{7.61}$$

and

$$K_c = \begin{bmatrix} 580 \times 10^6 & 145 \times 10^6 \\ 145 \times 10^6 & 290 \times 10^6 \end{bmatrix} - \begin{Bmatrix} 0 \\ 3.625 \times 10^6 \end{Bmatrix} \frac{1}{120{,}833} [0 \quad 3.625 \times 10^6]$$

$$= \begin{bmatrix} 580 \times 10^6 & 145 \times 10^6 \\ 145 \times 10^6 & 181.25 \times 10^6 \end{bmatrix} \tag{7.62}$$

Thus, equation (7.55), $F_{\text{condensed}} = K_{\text{condensed}} u_p$, becomes

$$\begin{Bmatrix} -1.1232 \times 10^6 \\ 2.0124 \times 10^6 \end{Bmatrix} = \begin{bmatrix} 580 \times 10^6 & 145 \times 10^6 \\ 145 \times 10^6 & 181.25 \times 10^6 \end{bmatrix} \begin{Bmatrix} u_4 \\ u_6 \end{Bmatrix} \tag{7.63}$$

Solving equation (7.63) for u_4 and u_6 we find

$$\begin{Bmatrix} u_4 \\ u_6 \end{Bmatrix} = \begin{Bmatrix} -.005890 \\ .015815 \end{Bmatrix} \tag{7.64}$$

From equation (7.52) we have

$$u_3 = \frac{1}{120{,}833}\left[-42120 - [0 \quad 3.625 \times 10^6]\begin{Bmatrix} -.005890 \\ .015815 \end{Bmatrix}\right] = -.82303 \tag{7.65}$$

The results are identical to Example 3.3.

Note that no information is lost in this process. Eventually, all displacements are obtained and all member forces can be determined. We have simply reduced the order of the matrices to be inverted.

7.7 AXIAL DEFORMATION IN FRAMES

You will recall from your structural analysis courses that axial deformation, as well as shear deformation, is generally neglected when solving frame problems using such techniques as slope-deflection and moment distribution. The neglect of these deformations can be justified by including them in some typical frames by using a solution technique such as virtual work, and then comparing their contributions to the total displacements with those of only the flexural terms. For the large majority of frame geometries the contributions are very small in comparison to those of bending.

We can formulate our structural stiffness equations in a way that neglects axial deformation by introducing constraint equations that will be developed in this section. However, before generating these equations formally, let us impose the necessary constraints directly on the frame shown in Figure 7-13. Equation (7.45), which is the reduced structural stiffness equation for this frame, is repeated here for convenience.

$$\begin{Bmatrix} 0 \\ 1 \\ 0 \\ 0 \\ 0 \\ 0 \\ 0 \\ 0 \end{Bmatrix} = \begin{bmatrix} 4EI/l & 6EI/l^2 & 0 & 2EI/l & 0 & 0 & 0 & 0 \\ - & 12EI/l^3 + EA/l & 0 & 6EI/l^2 & -EA/l & 0 & 0 & 0 \\ - & - & 3EI/l^3 + EA/l & 0 & 0 & -3EI/l^3 & 3EI/l^2 & 0 \\ - & - & - & 4EI/l & 0 & 0 & 0 & 0 \\ - & - & - & - & 12EI/l^3 + EA/l & 0 & 6EI/l^2 & 6EI/l^2 \\ - & sym. & - & - & - & 3EI/l^3 + EA/l & -3EI/l^2 & 0 \\ - & - & - & - & - & - & 7EI/l & 2EI/l \\ - & - & - & - & - & - & - & 4EI/l \end{bmatrix} \begin{Bmatrix} u_3 \\ u_4 \\ u_5 \\ u_6 \\ u_7 \\ u_8 \\ u_9 \\ u_{12} \end{Bmatrix} \tag{7.45}$$

If we neglect axial deformation of the two columns, then $u_5 = u_2 = 0$ and $u_8 = u_{11} = 0$, and rows 5 and 8 and columns 5 and 8 can be eliminated from equation (7.45).

Neglecting axial deformation in the horizontal girder implies that $u_7 = u_4$. We therefore add the elements in the column that multiplies u_7 to the column that multiplies u_4 and eliminate the u_7 column.

These operations yield

$$\begin{Bmatrix} F_3 \\ F_4 \\ F_6 \\ F_7 \\ F_9 \\ F_{12} \end{Bmatrix} = \begin{bmatrix} 4EI/l & 6EI/l^2 & 2EI/l & 0 & 0 \\ 6EI/l^2 & 12EI/l^3 & 6EI/l^2 & 0 & 0 \\ 2EI/l & 6EI/l^2 & 4EI/l & 0 & 0 \\ 0 & 12EI/l^3 & 0 & 6EI/l^2 & 6EI/l^2 \\ 0 & 6EI/l^2 & 0 & 7EI/l & 2EI/l \\ 0 & 6EI/l^2 & 0 & 2EI/l & 4EI/l \end{bmatrix} \begin{Bmatrix} u_3 \\ u_4 \\ u_6 \\ u_9 \\ u_{12} \end{Bmatrix} \tag{7.66}$$

The total horizontal force at the girder level is $F_4 + F_7$. We therefore take the row corresponding to F_7 and add its elements to the row corresponding to F_4. We obtain

$$\begin{Bmatrix} F_3 \\ F_4 + F_7 \\ F_6 \\ F_9 \\ F_{12} \end{Bmatrix} = \begin{bmatrix} 4EI/l & 6EI/l^2 & 2EI/l & 0 & 0 \\ 6EI/l^2 & 24EI/l^3 & 6EI/l^2 & 6EI/l^2 & 6EI/l^2 \\ 2EI/l & 6EI/l^2 & 4EI/l & 0 & 0 \\ 0 & 6EI/l^2 & 0 & 7EI/l & 2EI/l \\ 0 & 6EI/l^2 & 0 & 2EI/l & 4EI/l \end{bmatrix} \begin{Bmatrix} u_3 \\ u_4 \\ u_6 \\ u_9 \\ u_{12} \end{Bmatrix} \tag{7.67}$$

Equation (7.67) represents the structural stiffness equation for the frame when axial deformation is neglected. Solution for the displacements yields (with $F_4 + F_7 = P$),

$$\begin{Bmatrix} u_3 \\ u_4 \\ u_6 \\ u_9 \\ u_{12} \end{Bmatrix} = \frac{Pl^2}{EI} \begin{Bmatrix} -2/3 \\ 2l/3 \\ -2/3 \\ -1/3 \\ -5/6 \end{Bmatrix}$$

We now proceed to formalize the process of imposing constraints on certain displacements.

First, we partition the structural stiffness matrix as shown in equation (7.68).

$$\begin{Bmatrix} F_f \\ F_c \end{Bmatrix} = \begin{bmatrix} K_{ff} & K_{fc} \\ K_{cf} & K_{cc} \end{bmatrix} \begin{Bmatrix} u_f \\ u_c \end{Bmatrix} \tag{7.68}$$

In the above equation, the forces and displacements corresponding to the constraints to be imposed are subscripted with a "c." In the previous example, neglecting axial deformation,

$$\{F_f\} = \begin{Bmatrix} F_3 \\ F_4 \\ F_6 \\ F_9 \\ F_{12} \end{Bmatrix} \{F_c\} = \begin{Bmatrix} F_5 \\ F_7 \\ F_8 \end{Bmatrix} \{u_f\} = \begin{Bmatrix} u_3 \\ u_4 \\ u_6 \\ u_9 \\ u_{12} \end{Bmatrix} \{u_c\} = \begin{Bmatrix} u_5 \\ u_7 \\ u_8 \end{Bmatrix}$$

Expanding equation (7.68), we have

$$F_f = K_{ff}u_f + K_{fc}u_c \tag{7.69}$$

$$F_c = K_{cf}u_f + K_{cc}u_c \tag{7.70}$$

We next write a constraint equation that relates the constrained displacement to those that are unconstrained.

$$\{u_c\} = [C]\{u_f\} \tag{7.71}$$

or, for this example,

$$\begin{Bmatrix} u_5 \\ u_7 \\ u_8 \end{Bmatrix} = \begin{bmatrix} 0 & 0 & 0 & 0 & 0 \\ 0 & 1 & 0 & 0 & 0 \\ 0 & 0 & 0 & 0 & 0 \end{bmatrix} \begin{Bmatrix} u_3 \\ u_4 \\ u_6 \\ u_9 \\ u_{12} \end{Bmatrix} \tag{7.72}$$

Note that the above equation expresses the relationships $u_5 = 0$, $u_7 = u_4$, and $u_8 = 0$.

Using equation (7.71) in equations (7.69) and (7.70) we have

$$F_f = K_{ff}u_f + K_{fc}Cu_f \tag{7.73}$$

$$F_c = K_{cf}u_f + K_{cc}Cu_f \tag{7.74}$$

Of course, we could solve equation (7.73) for u_f; however, the matrix to be inverted, $K_{ff} + K_{fc}C$, is generally not symmetric. It would be more efficient from a numerical point of view to deal with a symmetric matrix. This can be accomplished by multiplying equation (7.74) by C^T and adding both equations.

Performing these operations we have

$$F_f + C^T F_c = (K_{ff} + K_{fc}C + C^T K_{cf} + C^T K_{cc}C)u_f \tag{7.75}$$

For this frame example, the left-hand side of equation (7.75) is

$$F_f + C^T F_c = \begin{Bmatrix} F_3 \\ F_4 \\ F_6 \\ F_9 \\ F_{12} \end{Bmatrix} + \begin{bmatrix} 0 & 0 & 0 \\ 0 & 1 & 0 \\ 0 & 0 & 0 \\ 0 & 0 & 0 \\ 0 & 0 & 0 \end{bmatrix} \begin{Bmatrix} F_5 \\ F_7 \\ F_8 \end{Bmatrix} = \begin{Bmatrix} F_3 \\ F_4 + F_7 \\ F_6 \\ F_9 \\ F_{12} \end{Bmatrix} \tag{7.76}$$

This is exactly the same as the left-hand side of equation (7.67).

After performing the operations indicated on the right-hand side of equation (7.75) we obtain the right-hand side of equation (7.67). Thus, equation (7.75) is identical to equation (7.67).

Approximate solutions can be obtained for problems where axial deformation is neglected by using a standard frame program and specifying large values for the cross-sectional areas of the members. Of course, this requires the inversion of a larger matrix. In the above example it would be necessary to invert an 8×8 rather than a 5×5 matrix. However, to obtain the smaller matrix there are a number of multiplications and additions to be performed, and the original reduced stiffness matrix must be reordered. Thus, computational efficiency is not as great as might be expected.

7.8 SUBSTRUCTURING

Often the designer is required to analyze a very large structure, perhaps involving thousands of nodes that are required to model the geometry accurately. Examples of this type of structure are large high-rise buildings, domes, and aircraft, to mention a few. Since there are very large numbers of degrees of freedom associated with the structure, computer facilities that are available may be inadequate to analyze the structure as a single unit. In addition, it is generally convenient to distribute the work involved in the project among several designers or design teams.

One procedure for accomplishing this is to divide the structure into smaller substructures, each of which can be assigned to a team for analysis.

Consider the frame shown in Figure 7-15.

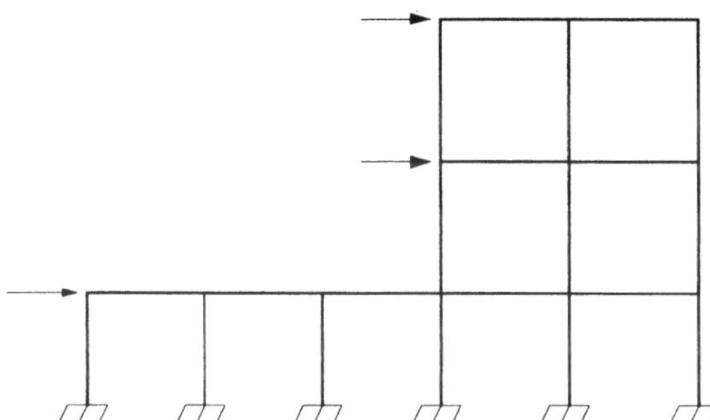

Figure 7-15 Two-dimensional frame.

For purposes of illustration, we divide this structure into three substructures as shown in Figure 7-16 where the connected nodes are emphasized.

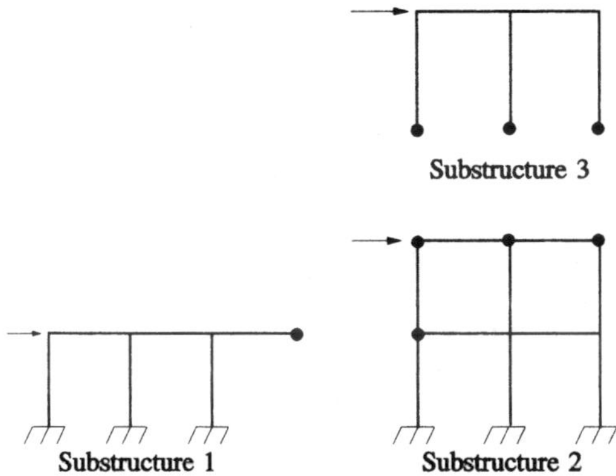

Figure 7-16 Substructures.

In the previous section, we used static condensation to reduce the order of our structural stiffness matrix. The effects of the degrees of freedom eliminated by this process were implicitly included in our final reduced set of equations.

Suppose we select the degrees of freedom at the interface nodes of each substructure as our primary quantities and apply the static condensation procedures to each substructure. The resulting condensed stiffness matrices will be much smaller than the stiffness matrices obtained considering all non-zero degrees of freedom. For example, the stiffness matrix for substructure 1 will reduce in size from a 12×12 to a 3×3 matrix. Similarly, substructures 2 and 3 reduce from an 18×18 to a 12×12 and from an 18×18 to a 9×9, respectively.

We can then combine the reduced force and stiffness matrices as before by enforcing equilibrium at the interface nodes. In effect, we have reduced the original structure to three large elements. For this example, if we treated the original structure as a single unit, we would have a 36×36 matrix to invert. Using the substructuring procedure described, we have a 12×12 matrix to invert in order to find the displacements at the interface nodes.

After finding these displacements, the force-displacement relationships for each substructure (equation [7.52]) are used to determine the interior displacements. Member forces are found as before.

Example 7.7

Consider the frame shown in Figure 7-17. We wish to analyze the frame by substructuring.

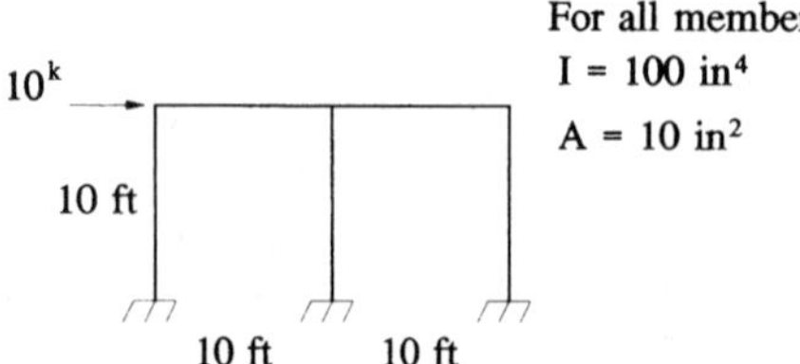

Figure 7-17 Example 7.7.

We shall divide the structure into two substructures as shown in Figure 7-18. The connecting node is emphasized.

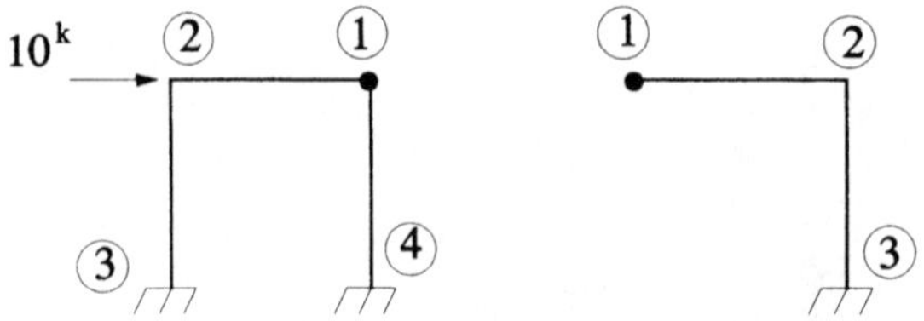

Figure 7-18 Division into substructures.

The force-displacement relationship for substructure 1 is

$$\begin{Bmatrix} 0 \\ 0 \\ 0 \\ 10 \\ 0 \\ 0 \end{Bmatrix} = \begin{bmatrix} 2436.81 & 0 & 1208.33 & -2416.67 & 0 & 0 \\ 0 & 2436.81 & -1208.33 & 0 & -20.14 & -1208.33 \\ 1208.33 & -1208.33 & 193333 & 0 & 1208.33 & 48333 \\ -2416.67 & 0 & 0 & 2436.81 & 0 & 1208.33 \\ 0 & -20.14 & 1208.33 & 0 & 2436.81 & 1208.33 \\ 0 & -1208.33 & 48333 & 1208.33 & 1208.33 & 193333 \end{bmatrix} \begin{Bmatrix} u_1 \\ u_2 \\ u_3 \\ u_4 \\ u_5 \\ u_6 \end{Bmatrix} \tag{7.77}$$

We want to use u_1, u_2, and u_3 as primary quantities. In this case the stiffness matrix in equation (7.66) is partioned into four 3×3 matrices in accordance with equation (7.49). The force matrix is partioned into two 3×1 matrices.

Performing the operations indicated by equation (7.54) we find for substructure 1

$$\begin{Bmatrix} 9.95 \\ -.031 \\ 1.232 \end{Bmatrix} = \begin{bmatrix} 32.71 & 7.47 & 910.6 \\ 7.47 & 2429.2 & -900.6 \\ 910.6 & -900.6 & 180{,}875 \end{bmatrix} \begin{Bmatrix} u_1 \\ u_2 \\ u_3 \end{Bmatrix} \tag{7.78}$$

For substructure 2 the force-displacement relationship is

$$\begin{Bmatrix} 0 \\ 0 \\ 0 \\ 0 \\ 0 \\ 0 \end{Bmatrix} = \begin{bmatrix} 2416.67 & 0 & 0 & -2416.67 & 0 & 0 \\ 0 & 20.14 & 1208.33 & 0 & -20.14 & 1208.33 \\ 0 & 1208.33 & 96666 & 0 & -1208.33 & 48333 \\ -2416.67 & 0 & 0 & 2436.81 & 0 & 1208.33 \\ 0 & -20.14 & -1208.33 & 0 & 2436.81 & -1208.33 \\ 0 & 1208.33 & 48333 & 1208.33 & -1208.33 & 193333 \end{bmatrix} \begin{Bmatrix} u_1 \\ u_2 \\ u_3 \\ u_4 \\ u_5 \\ u_6 \end{Bmatrix} \tag{7.79}$$

As in the case of substructure 1, the partitioning necessary to retain u_1, u_2, and u_3 as primary quantities results in 3×3 and 3×1 matrices.

Equation (7.54) yields

$$\begin{Bmatrix} 0 \\ 0 \\ 0 \end{Bmatrix} = \begin{bmatrix} 12.50 & -7.47 & -297.7 \\ -7.47 & 12.50 & 900.6 \\ -297.7 & 900.6 & 84208.4 \end{bmatrix} \begin{Bmatrix} u_1 \\ u_2 \\ u_3 \end{Bmatrix} \tag{7.80}$$

Combining equations (7.78) and (7.80) in the usual fashion we obtain

$$\begin{Bmatrix} 9.95 \\ -.031 \\ 1.232 \end{Bmatrix} = \begin{bmatrix} 45.2 & 0 & 612.9 \\ 0 & 2441.7 & 0 \\ 612.9 & 0 & 265083.4 \end{bmatrix} \begin{Bmatrix} u_1 \\ u_2 \\ u_3 \end{Bmatrix} \tag{7.81}$$

Solving equation (7.81) for the displacements we obtain:

$$\begin{Bmatrix} u_1 \\ u_2 \\ u_3 \end{Bmatrix} = \begin{Bmatrix} .2275\ in \\ -.000013\ in \\ -.00052\ rad \end{Bmatrix} \tag{7.82}$$

We next use equation (7.52) for each substructure to determine the remaining displacements.

For substructure 1:

$$\begin{Bmatrix} u_4 \\ u_5 \\ u_6 \end{Bmatrix} = \begin{Bmatrix} .2221\ in \\ .00088\ in \\ -.001264\ rad \end{Bmatrix} \tag{7.83}$$

For substructure 2:

$$\begin{Bmatrix} u_4 \\ u_5 \\ u_6 \end{Bmatrix} = \begin{Bmatrix} .2263\ in \\ -.00090\ in \\ -.00129\ rad \end{Bmatrix} \tag{7.84}$$

Note that the displacement subscripts refer to the individual substructure nodal numbers indicated in Figure 7-18.

7.9 NON-UNIFORM MEMBERS

There are some structures whose analysis could benefit by modeling them using elements other than those considered up to this point in the text. Some of these include curved members and beams where shear deformation is important. Others include tapered bars or beams, haunched members, and one-way slabs with drop panels as shown in Figure 7-19.

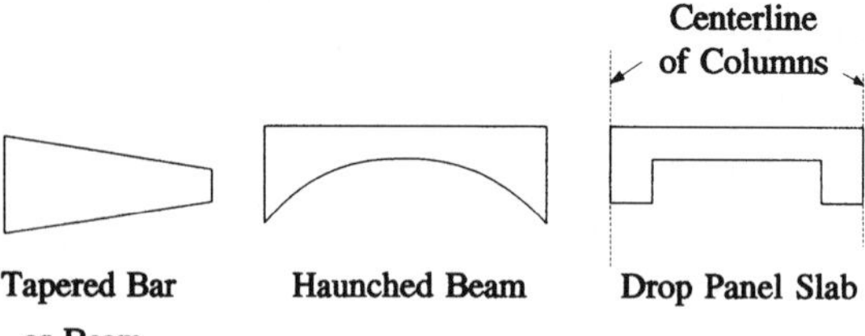

Figure 7-19 Non-uniform members.

We can model the members shown in Figure 7-19 with a number of uniform cross-sectional members, although doing this results in an approximate solution. In fact, one of the problems in Chapter 1 dealt with modeling a tapered one-dimensional rod element with a varying number of uniform elements. Naturally, the storage requirements and time of solution will increase owing to the larger number of members required to reasonably model the structure and the resultant increase in the number of degrees of freedom. For smaller structures, the trade-off between the use of a simpler computer program and an increase in storage requirements and solution time is generally acceptable. However, if very accurate solutions are required, or if it is necessary to analyze very large structures or ones that incorporate many special elements, then it is more efficient to develop the force-displacement relationships for these less common elements.

7.9a Linearly Tapered Bar

As one example of a special element, consider the linearly tapered one-dimensional bar shown in Figure 7-20. We want to determine the stiffness matrix for this element.

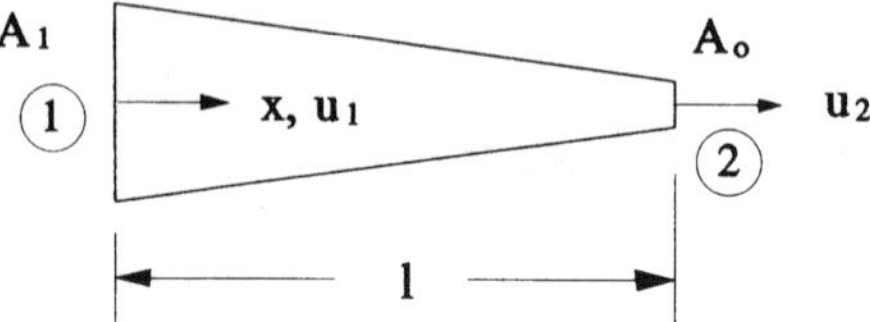

Figure 7-20 Linearly tapered bar.

As in the case of the uniform rod, there are two degrees of freedom that will result in a 2 × 2 stiffness matrix.

The cross-sectional area of the rod as a function of position x is given by

$$A(x) = A_1 + \left(\frac{A_o - A_1}{l}\right) x \tag{7.85}$$

We fix node 1, apply a force P to node 2, and determine the displacement of node 2.

The unit strain at any point in the rod is P/AE, where P and E are constant and A is given by equation (7.85). The integral of this strain taken over the length of the rod yields the displacement at node 2, u_2. We have

$$u_2 = u(l) = \int_0^l \epsilon dx = \frac{P}{E}\int_0^l \frac{dx}{A_1 + \left(\frac{A_0 - A_1}{l}\right)x} \tag{7.86}$$

After integrating we find

$$u_2 = \frac{P}{E}\left(\frac{l}{A_o - A_1}\right)\ln\frac{A_o}{A_1} \tag{7.87}$$

Solving for the force P due to a unit displacement of u_2 yields the stiffness coefficient k_{22}.

$$k_{22} = \frac{E(A_o - A_1)}{l\ln\left(\frac{A_o}{A_1}\right)} \tag{7.88}$$

The force at node 1 corresponding to this displacement is equal to the negative of the previous value, thus $k_{12} = -k_{22}$.

Fixing node 2 and applying a load P to node 1 will yield $u_1 = u_2 = -k_{12} = -k_{21}$. Thus, our force-displacement relationship for the linearly tapered rod becomes

$$\begin{Bmatrix} F_1 \\ F_2 \end{Bmatrix} = \frac{E(A_o - A_1)}{l\ln\left(\frac{A_o}{A_1}\right)}\begin{bmatrix} 1 & -1 \\ -1 & 1 \end{bmatrix}\begin{Bmatrix} u_1 \\ u_2 \end{Bmatrix} \tag{7.89}$$

Keep in mind that for more complicated cross-sectional variations it may be necessary to resort to numerical integration of equation (7.86).

7.9b Stepped bar

Consider the stepped bar shown in Figure 7-21.

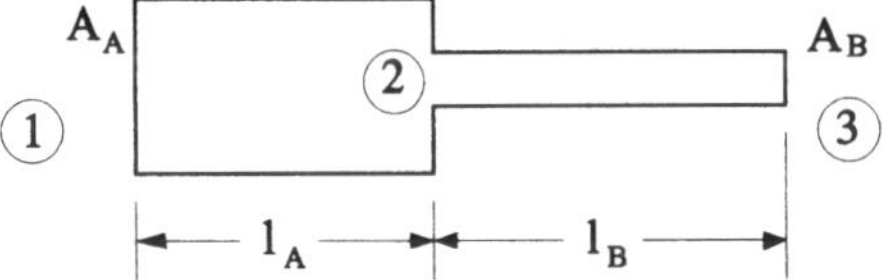

Figure 7-21 Stepped bar.

Naturally, we could treat this member as two separate elements, each having a constant cross-sectional area. Suppose, however, that we wish to determine the force-displacement relationship for the entire member. That is, we want to generate the equation

$$\begin{Bmatrix} F_1 \\ F_3 \end{Bmatrix} = [k]\begin{Bmatrix} u_1 \\ u_3 \end{Bmatrix} \tag{7.90}$$

Although we could use the same procedure as we used for the tapered bar—that is, fix node 1, apply a force to node 3, and determine the resulting displacement—we choose to use static condensation to illustrate an alternative approach when sections of the element have constant areas.

Designating EA_A/l_A and EA_B/l_B as k_A and k_B respectively, and treating u_2 as the secondary quantity, the overall structural stiffness equation becomes

$$\begin{Bmatrix} F_1 \\ F_3 \\ F_2 \end{Bmatrix} = \begin{bmatrix} k_A & 0 & -k_A \\ 0 & k_B & -k_B \\ -k_A & -k_B & k_A + k_B \end{bmatrix} \begin{Bmatrix} u_1 \\ u_3 \\ u_2 \end{Bmatrix} \tag{7.91}$$

Note that these equations have been rearranged to conform with equation (7.49). Equation (7.54) yields the required force-displacement relationship:

$$\begin{Bmatrix} F_1 \\ F_3 \end{Bmatrix} = \frac{k_A k_B}{k_A + k_B} \begin{bmatrix} 1 & -1 \\ -1 & 1 \end{bmatrix} \begin{Bmatrix} u_1 \\ u_3 \end{Bmatrix} \tag{7.92}$$

This is simply the combination of stiffnesses of two axial springs connected in series. Notice, however, that we have generated a "superelement." This is exactly what was accomplished by substructuring in section 7.8. This produces an exact representation of the behavior of the member.

7.9c Linearly Tapered Beam

Consider a constant width, linearly tapered beam as shown in Figure 7-22.

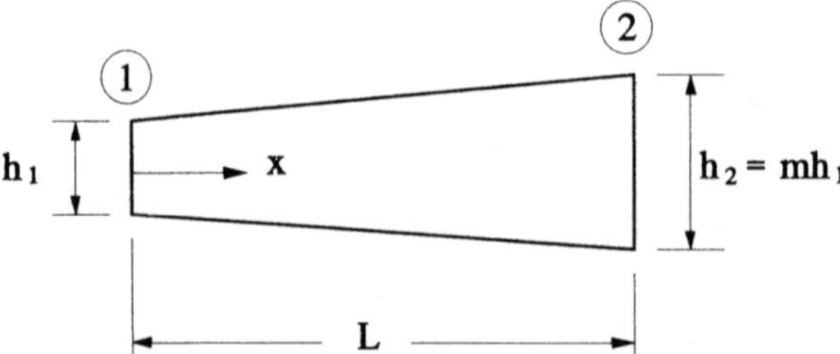

Figure 7-22 Linearly tapered beam.

The depth of the beam can be written

$$h(x) = h_1 \left[1 + (m-1)\frac{x}{L}\right] \tag{7.93}$$

Letting $n = m - 1$, the moment of inertia becomes

$$I = \frac{1}{12} b h_1^3 \left[1 + n\frac{x}{L}\right]^3 \tag{7.94}$$

To illustrate one procedure that can be used, we will determine the stiffness elements k_{11} and k_{21} by using the basic definition of k_{ij}.

We introduce a vertical deflection at node 1 keeping all other displacements zero as shown in Figure 7-23.

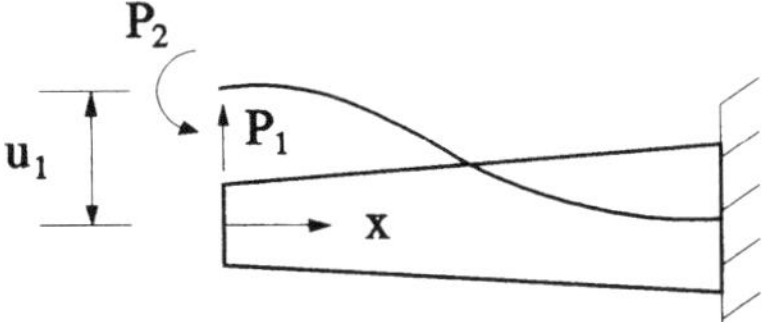

Figure 7-23 Vertical deflection at node 1.

The moment-curvature relationship can be written

$$EI\frac{d^2y}{dx^2} = P_1x - P_2 \tag{7.95}$$

Using equation (7.94) for I and integrating once we find

$$\frac{dy}{dx} = \frac{P_2L}{2EI_1n}\left[1+\frac{nx}{L}\right]^{-2} + \frac{P_1}{EI_1}\frac{\left(1+\frac{nx}{L}\right)^{-2}}{n^2/L^2}\left[\frac{1}{2}-\left(1+\frac{nx}{L}\right)\right] + C_1 \tag{7.96}$$

Integrating once again, we have

$$y = -\frac{P_2L^2}{2EI_1n^2}\left(1+\frac{nx}{L}\right)^{-1} - \frac{P_1}{2EI_1n^3/L^3}\left(1+\frac{nx}{L}\right)^{-1}$$
$$-\frac{P_1}{EI_1n^3/L^3}\ln\left(1+\frac{nx}{L}\right) + C_1x + C_2 \tag{7.97}$$

The constants of integration C_1 and C_2 are found by applying the boundary conditions $y'(L) = 0$ and $y(L) = 0$. We obtain

$$C_1 = -\frac{P_2L}{2EI_1n}(1+n)^{-2} + \frac{P_1}{EI_1}\frac{(1+n)^{-2}}{n^2/L^2}\left[\frac{1}{2}+n\right] \tag{7.98}$$

$$C_2 = \frac{P_2L^2}{2EI_1n^2}(1+n)^{-1} + \frac{P_1}{2EI_1n^3/L^3}(1+n)^{-1} + \frac{P_1}{EI_1n^3/L^3}\ln(1+n) - C_1L \tag{7.99}$$

We next determine the forces P_1 and P_2 by using the boundary conditions at node 1; namely, $y'(0) = 0$ and $y(0) = u_1 = 1$.

These forces are those at node 1 corresponding to a unit vertical displacement of node 1. They are, therefore, k_{11} and k_{21}.

To illustrate, let $m = 2$. Then $n = 1$.

Evaluating C_1 and C_2 (equations [7.98] and [7.99]), applying the boundary conditions at node 1, and solving for P_1 and P_2, we obtain

$$P_1 = k_{11} = 37.76\,\frac{EI_1}{L^3} \tag{7.100}$$

$$P_2 = k_{21} = 12.59\,\frac{EI_1}{L^2} \tag{7.101}$$

The other stiffness elements are obtained in a similar way.

We will develop another method for obtaining elemental stiffnesses in Chapter 8.

7.10 SUMMARY

In this chapter we investigated several additional topics in the stiffness formulation of matrix structural analysis. These included a discussion of bandwidth, combination of different elements, elastic supports, inclined supports, internal hinges, static condensation, axial deformation in frames, substructuring, and non-prismatic members.

Some of the techniques developed are used often and others less frequently. However, all contribute to our ability to analyze complex structures by the matrix method.

PROBLEMS

Determine the bandwidth for the problems indicated below. Try to minimize the bandwidth by efficient node numbering.

7.1 Problem 2.2

7.2 Problem 2.3

7.3 Problem 2.11

7.4 Problem 3.17

7.5 Problem 3.18

7.6 Problem 3.20

7.7 Problem 4.5

7.8 Problem 4.6

7.9 Problem 5.1

7.10 Problem 5.3

7.11 Problem 5.6

7.12 Problem 6.1

7.13 Problem 6.2

7.14 Problem 6.6

7.15 Problem 6.8

7.16 Generate the reduced structural stiffness matrix for the frame shown in Figure P7-16. For the frame members, $E = 29 \times 10^6$ psi, $I = 100$ in^4, and $A = 10$ in^2. For the truss member, $E = 29 \times 10^6$ psi, and $A = 2$ in^2.

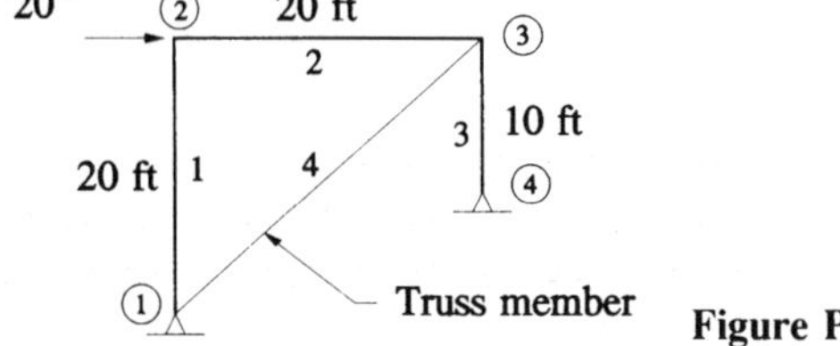

Figure P7-16

7.17 Generate the reduced stiffness matrix for the frame shown in Figure P7-17. For the frame members, $E = 29 \times 10^6$psi, $I = 120$ in^4, and $A = 2$ in^2. For the truss members, $E = 29 \times 10^6$ psi, and $A = 1$ in^2.

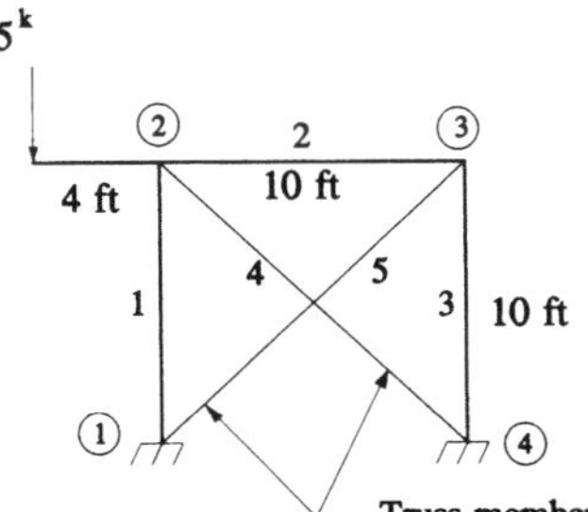

Figure P7-17

7.18 Verify the alternative solution to Example 7.2.

7.19 Solve for nodal displacements and member forces of the truss shown in Figure P7-19. For all members, $E = 29 \times 10^6$ psi, and $A = 2$ in^2.

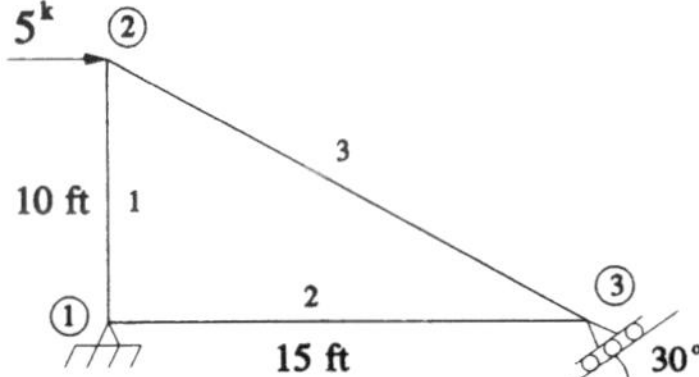

Figure P7-19

7.20 Solve problem 7.19 by placing a very stiff member perpendicular to the inclined support in Figure P7-19. Compare your results with those of problem 7.19.

7.21 Generate the reduced stiffness matrix for the frame shown in Figure P7-21 by

(a) considering the hinge to be at the left end of member 2;

(b) considering the hinge to be at the right end of member 1

($E = 29 \times 10^6$ psi, $A = 10$ in^2, and $I = 100$ in^4 for both members).

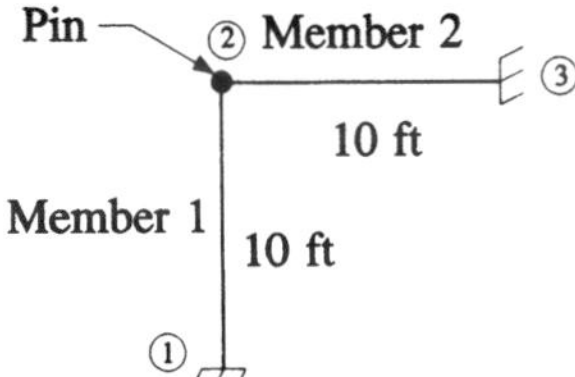

Figure P7-21

7.22 Consider the stepped beam shown in Figure P7-22.

(a) Treating the beam as two members, generate the overall stiffness equation.

(b) Treating the displacements at node 2 as secondary quantities, use static condensation to generate a set of force-displacement relationships. $E = 29 \times 10^6$ psi.

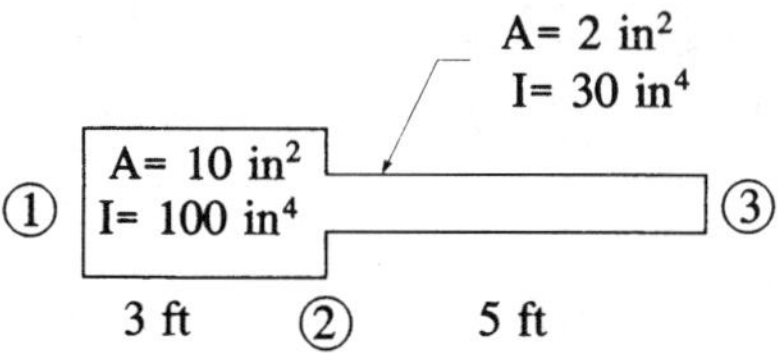

Figure P7-22

7.23 Use the method of substructuring to analyze the truss shown in Figure P7-23.

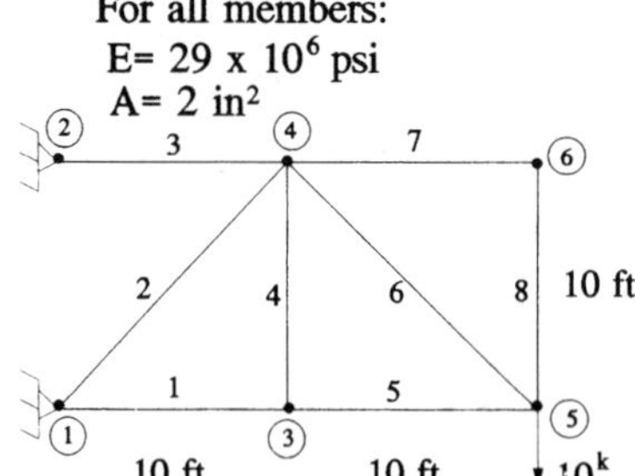

Figure P7-23

7.24 Use the method of substructuring to analyze the frame shown in Figure P7-24.

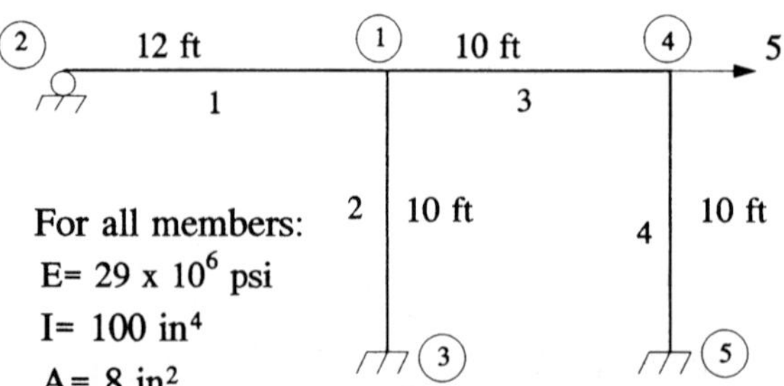

Figure P7-24

7.25 Analyze the linearly tapered bar shown in Figure P7-25 by

(a) treating it as a single element

(b) treating it as two elements of equal length.

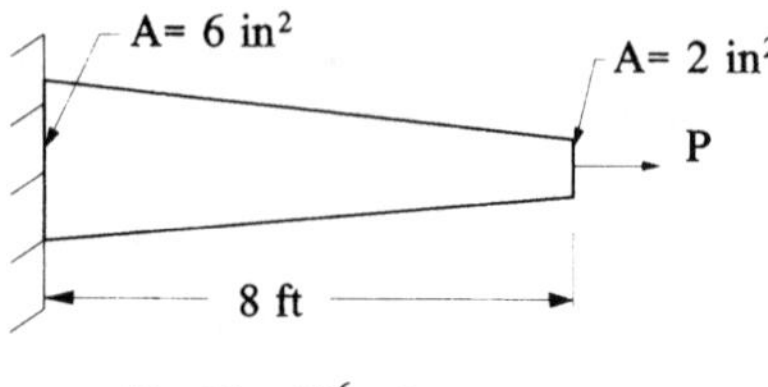

E = 29 x 10⁶ psi

Figure P7-25

7.26 Determine the elements k_{22} and k_{12} for the linearly tapered beam shown in Figure 7-22. Use $m = 2$.

CHAPTER 8

VIRTUAL WORK AND THE PRINCIPLE OF MINIMUM POTENTIAL ENERGY

8.1 INTRODUCTION

Up to this point in the text we have used the relationships of basic strength of materials and elementary structural analysis to derive elemental stiffnesses and to determine the effects of non-nodal loads. Equilibrium considerations were then used to generate the global structural stiffness equations.

In this chapter we present alternative methods of deriving elemental stiffnesses, equivalent nodal forces, and the structural stiffness equation. The techniques we shall use are based on work and energy principles. We shall also find that the techniques presented are very useful in finding approximate solutions to structural problems.

When it is desired to formulate stiffnesses for more complicated elements such as plate and shell elements, in most cases we must resort to these alternative work and energy methods since basic structural theory does not allow us to use a direct method; namely exact force-displacement relationships for these elements are not known. For example, suppose we wanted to formulate the stiffness matrix for a triangular plate element loaded in its plane. Fig. 8-1 shows such an element.

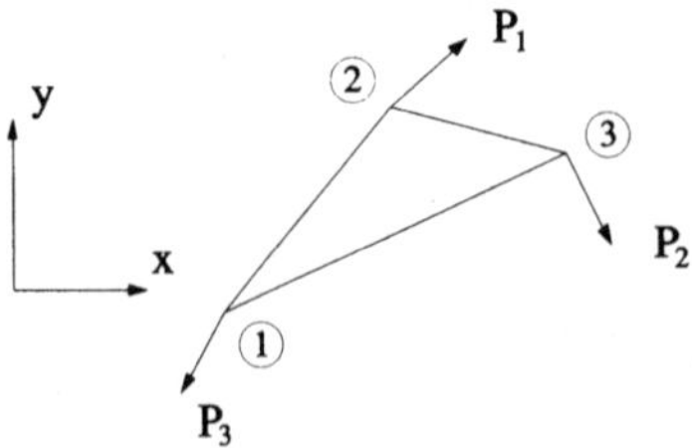

Figure 8-1 Triangular plate element.

In Fig. 8-1, we have shown a three-node triangular plate element. Since this element lies in the x-y plane, we have two possible translations in the coordinate directions at each node. Thus, this element has six degrees of freedom.

To generate the stiffness matrix from the basic definition of k_{ij}, we would have to determine the force-displacement relationships by individually introducing six unit displacements and finding the nodal forces corresponding to each of these displacements. Basic structural theory does not allow us to do this directly, and we therefore must resort to other techniques. These techniques are generally based on work and energy methods. This type of element is typically treated in a study of finite elements, and this chapter should give you a beginning background for a following course in the finite-element method.

8.2 THE PRINCIPLE OF VIRTUAL WORK

8.2a Rigid Bodies

Consider a rigid body in equilibrium under a set of Q forces as shown in Figure 8-2.

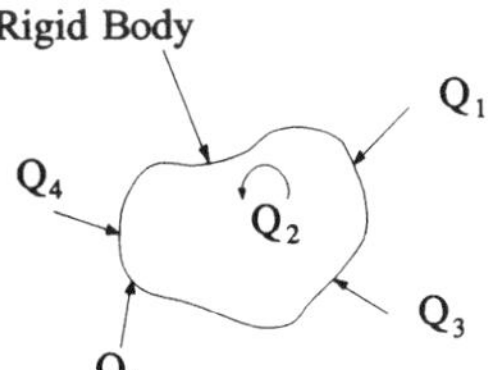

Figure 8-2 Rigid body in equilibrium.

Note that the Q system is an equilibrium force system and therefore includes the reactions as well as applied forces.

Suppose some external effect causes a displacement of this rigid body. For small displacements we can represent this displacement as a sum of a translation and a rotation. Note that the Q-force system will do work acting through the imposed displacement. This displacement is called a "virtual displacement" and the work "virtual work." Let us now calculate the virtual work done by the Q forces acting through the translation (Figure 8-3).

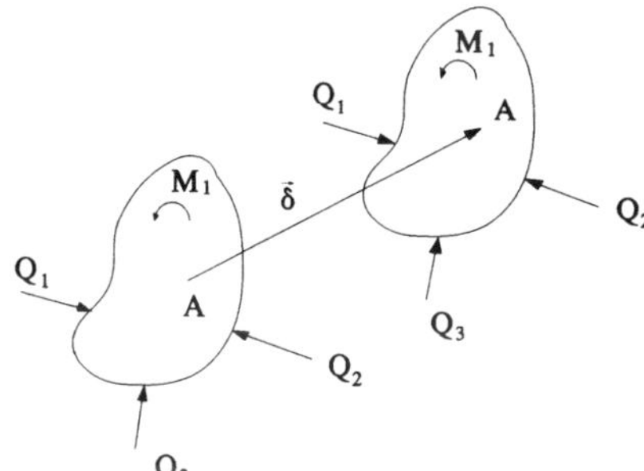

Figure 8-3 Translation of rigid body.

Since the Q forces are at full value when the translation occurs, and since the work of a constant force acting through a vector translation δ is the dot product of the force and the translation, we have, for N forces

$$\textit{Virtual work in translation} = \vec{\mathbf{Q}}_1 \cdot \vec{\delta} + \vec{\mathbf{Q}}_2 \cdot \vec{\delta} + \ldots + \vec{\mathbf{Q}}_N \cdot \vec{\delta} = \left(\sum_{i=1}^{N} \vec{\mathbf{Q}}_i \right) \cdot \vec{\delta} \quad (8.1)$$

since δ is the same for all forces.

Remember, however, that the Q system is an equilibrium force system, thus the sum of the Q forces is zero. We can therefore state:

The work done by an equilibrium force system acting through a virtual translation is zero.

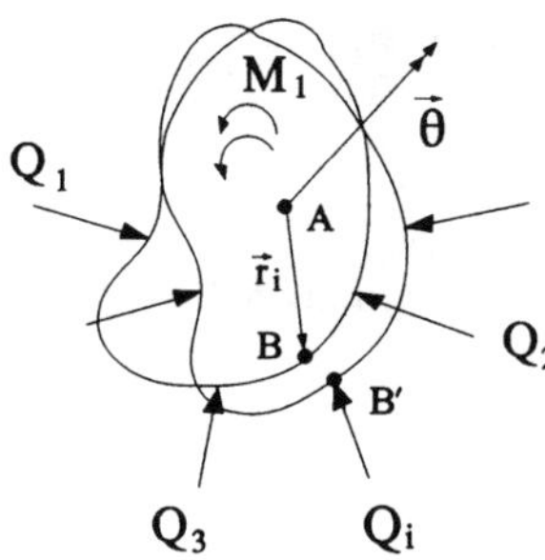

Figure 8-4 Rotation of rigid body.

Next consider a small rotation θ.

In Figure 8-4, the position vector from point A to point B, which is the point of application of the force $\vec{\mathbf{Q}}_i$, is designated as $\vec{\mathbf{r}}_i$. Of course, moments do work acting through rotations, thus

$$\text{Work done by } \vec{\mathbf{Q}}_i = (\vec{\mathbf{r}}_i \times \vec{\mathbf{Q}}_i) \cdot \vec{\theta} \tag{8.2}$$

For any applied moments $\vec{\mathbf{M}}_j$

$$\text{Work done by } \vec{\mathbf{M}}_j = \vec{\mathbf{M}}_j \cdot \vec{\theta} \tag{8.3}$$

Adding the work from all applied forces and moments we have

$$\begin{aligned} \textit{Virtual work through rotation} &= \sum_{i=1}^{N} (\vec{\mathbf{r}}_i \times \vec{\mathbf{Q}}_i) \cdot \vec{\theta} + \sum_{j=1}^{M} \vec{\mathbf{M}}_j \cdot \vec{\theta} \\ &= \left\{ \sum_{i=1}^{N} (\vec{r}_i \times \vec{\mathbf{Q}}_i) + \sum_{j=1}^{K} \vec{\mathbf{M}}_j \right\} \cdot \vec{\theta} \end{aligned} \tag{8.4}$$

where K is the number of applied moments in the Q system.

The term in parentheses is simply a moment equilibrium equation and must equal zero since the Q system is in equilibrium. Thus,

The work done by an equilibrium force system acting through a virtual rotation is zero.

The principle of virtual work for rigid bodies can therefore be stated:

The total work done by an equilibrium force system acting through a virtual displacement is zero.

This principle is often used to find reactions and internal forces for any determinate structure. As an example, consider the beam shown in Figure 8-5.

Since the applied loads cause very small elastic displacements of the beam, we shall consider the structure as a rigid body. We replace the supports with the reactions corresponding to them.

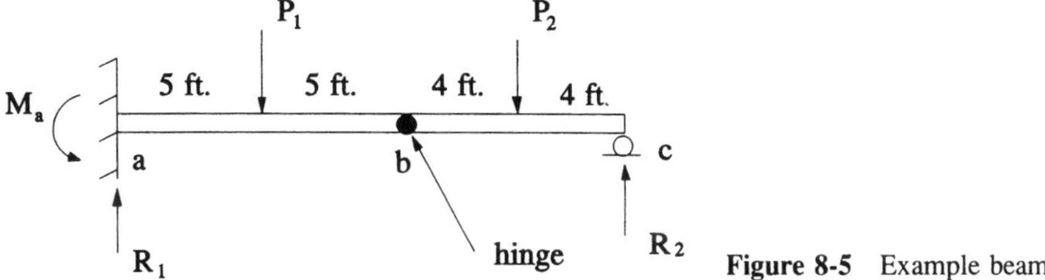

Figure 8-5 Example beam.

To determine R_2, introduce the virtual displacement shown in Figure 8-6.

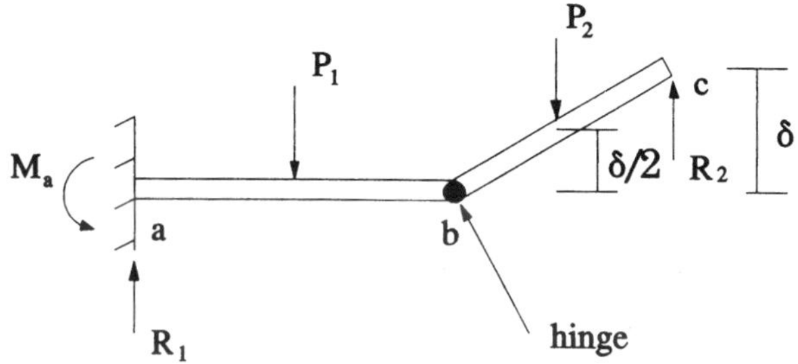

Figure 8-6 Virtual displacement for finding R_2.

Next, we calculate the virtual work and equate it to zero.

$$R_2\delta - P_2\delta/2 = 0 \tag{8.5}$$

$$R_2 = P_2/2 \tag{8.6}$$

To find M_a, use the virtual displacement shown in Figure 8-7.

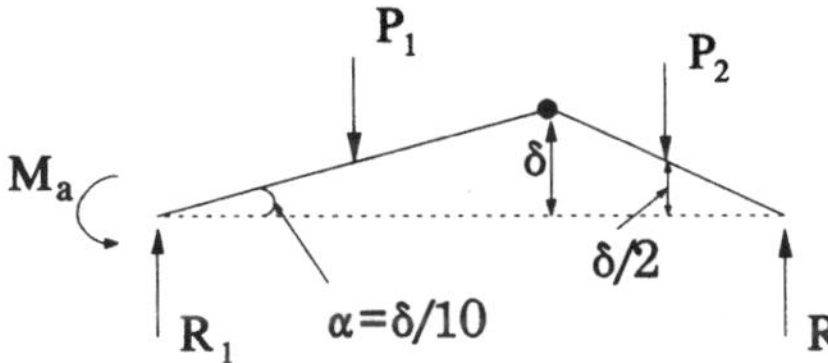

Figure 8-7 Virtual displacement for finding M_a.

For small displacements, α is approximately equal to tan $\alpha = \delta/10$.

$$\text{Virtual work} = 0 = M_a\delta/10 - P_1\delta/2 - P_2\delta/2 \tag{8.7}$$

$$M_a = 10\ (P_1/2 + P_2/2) \tag{8.8}$$

For R_1, introduce the virtual displacement shown in Figure 8-8.

$$\text{Virtual work} = R_1\delta - P_1\delta - P_2\delta/2 = 0 \tag{8.9}$$

$$R_1 = P_1 + P_2/2 \tag{8.10}$$

Note that we have introduced virtual displacements that allow only one unknown force at a time to do work. This is not necessary, but it avoids having to solve simultaneous equations. Keep in mind that we have actually written equilibrium equations but

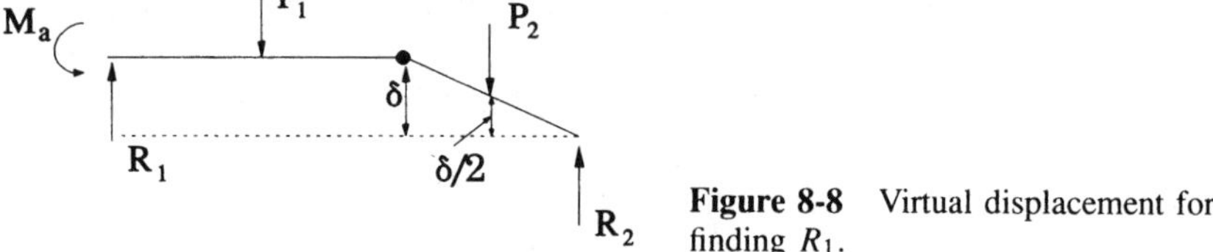

Figure 8-8 Virtual displacement for finding R_1.

used virtual work to do it. Equation (8.5), for example, is simply a moment equilibrium equation written about point b for member bc as a free body. We next consider the principle of virtual work for deformable bodies.

8.2b Deformable Bodies

We shall now use the principle of virtual work for rigid bodies to develop the method of virtual work for deformable bodies.

Consider a deformable body in equilibrium under a set of Q forces. As in the case of a rigid body, this set of forces includes reactions as well as applied forces. These Q forces develop stresses throughout the body. Since the entire body is in equilibrium, then any portion of the body we wish to consider in a free-body diagram must also be in equilibrium. Consider the small elements shown in Figure 8-9a.

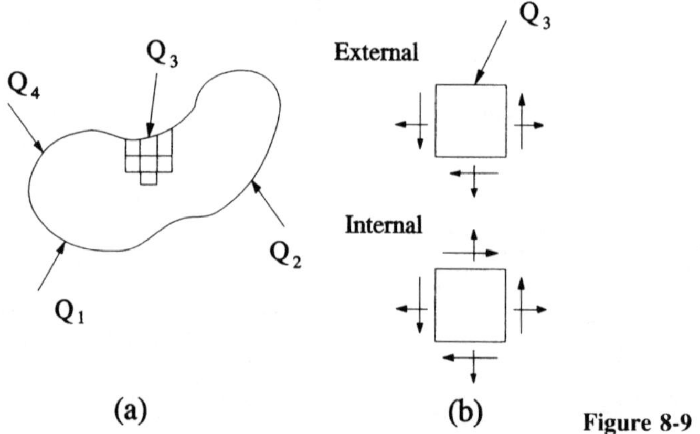

Figure 8-9 Elastic body.

Two types of elements are present as shown in Figure 8-9b: (1) internal elements that have faces in common with adjacent elements, and (2) elements that have external boundaries that have Q forces acting on them. On faces of internal elements, there are stresses on common faces equal in magnitude but opposite in direction to those on the adjacent element.

The body is now subjected to a distortion from some external P effect, a "virtual distortion." The elements of the body will, in general, be distorted as well as displaced as a rigid body. Since the element faces will displace, the stresses acting on these faces will do work. Let the total work done by the Q forces and Q-induced stresses on all elements be designated by W_{stress}. Part of this work is due to the distortion of the element faces. Call this W_{d}. The balance of the total work, $W_{\text{stress}} - W_{\text{d}}$, is due to work done by the stresses acting through a rigid body motion. We have seen, however, that work done by an equilibrium system acting through a rigid body motion is zero. Thus,

$$W_{\text{stress}} = W_{\text{d}} \tag{8.11}$$

Remember that W_{stress} represents the total work done by the boundary stresses and external boundary loads acting on the elements of the body. On internal element boundaries, equal and opposite stresses act on common faces of adjacent elements. Since the face is common and undergoes some distortion, the equal but opposite stresses do equal but opposite work. Thus, the net work done by internal boundary stresses is zero. Therefore the total work W_{stress} is equal only to the work done by external forces that act on external element boundaries. We call this "external virtual work."

In equation (8.11), W_{d} is the work done by the Q-induced stresses acting through the distortion caused by some external P effect. This is called "internal virtual work."

We can now state the principle of virtual work for deformable bodies:

The external work done by the Q equilibrium force system acting through the virtual displacements induced by the P system is equal to the internal virtual work done by the Q-induced stresses acting through the admissible virtual displacements caused by the P effect.

We have used the term "admissible" in the statement of the principle of virtual work for deformable bodies. This means that the P effect cannot create displacements of the elastic body which would violate support conditions or create discontinuities of displacements in the body.

Thus far we have not been concerned about the form of the P effect that causes the distortion of the elastic body. The P effect could be, but is not limited to, a second force system. It could also be a system involving temperature changes, incorrect bar lengths, etc. Recall from your basic structural analysis courses that the principle of virtual work was used in the formulation of the "dummy unit load" method for finding displacements of a structure.

Since external virtual work is equal to the work done by the external Q_{n} forces acting through the imposed displacements, the general expression for this work is

$$\text{External virtual work} = \sum_{n=1}^{N} Q_n \delta_n^P \tag{8.12}$$

where δ_n^P is the displacement at point n corresponding to Q_{n} but caused by the P effect. That is, if Q_n is a linear force, then δ_n^P is the displacement at point n along the line of action of Q_n. If Q_n is a couple, then δ_n^P is a vector rotation in the same direction as the moment vector Q_n.

The internal virtual work (IVW) is the work done by the Q-induced stresses acting through the displacement imposed and will depend on the type of stress and the specific P effect causing the displacement. We now derive the expressions for internal virtual work for axial stress and bending stress.

8.2c Expressions for Internal Virtual Work

Consider the case when the Q-force system results in axial stresses in the body. Naturally, a truss is an example of this type of structure.

Figure 8-10 shows a member subjected to a Q-force-system-induced axial stress. Also shown is the displacement corresponding to the axial *force* induced by the P effect.

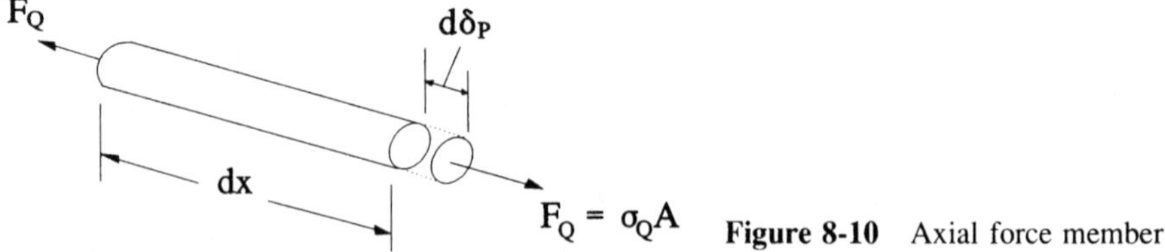

Figure 8-10 Axial force member.

The internal virtual work for this element is $F_Q d\delta_P$. Suppose that $d\delta_P$ is due to a second system of forces. Then for this element,

$$d\delta_P = \frac{F_P}{EA}dx \tag{8.13}$$

For the entire bar,

$$\delta_P = \int_0^l \frac{F_P F_Q}{EA}dx \tag{8.14}$$

For the entire truss,

$$Total\ IVW = \sum_{i=1}^{N}\left[\int_0^l \frac{F_Q^i F_P^i}{EA}dx\right] \tag{8.15}$$

where N is the total number of axial force members in the truss, and i represents the ith member.

If all members have constant forces and areas throughout their lengths, equation (8.15) becomes

$$IVW = \sum_{i=1}^{N} \frac{F_Q^i F_P^i l_i}{EA} \tag{8.16}$$

If δ_P was due to a temperature change of one or more members, then

$$\delta^i_P = \alpha_i(\Delta T)_i l_i \tag{8.17}$$

and

$$IVW = \sum_{i=1}^{K} F^i_Q \alpha_i (\Delta T)_i l_i \tag{8.18}$$

where the summation is taken over K members having temperature changes.

If δ_P was due to an incorrect bar length of one or more members, say Δ_i, then

$$IVW = \sum_{i=1}^{M} F^i_Q \Delta_i \tag{8.19}$$

where M represents the number of members with incorrect lengths.

Of course, since we are dealing with a linear-elastic system, superposition is valid and the effects of all of these P systems can be added.

Next consider the case where both the Q and P systems are load systems that cause bending stresses and displacements in a beam or frame element. Figure 8-11 depicts this case.

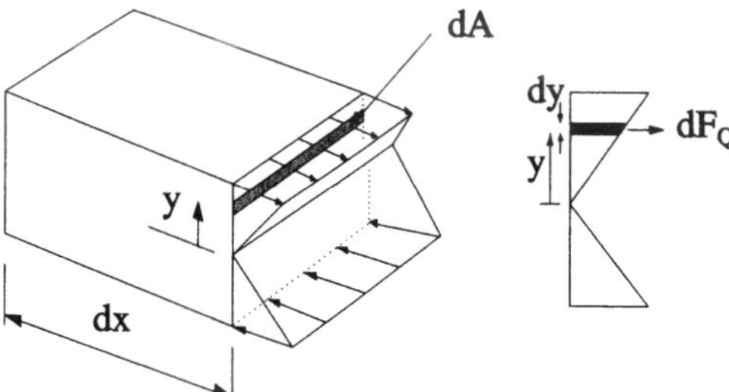

Figure 8-11 Bending stress.

Now,

$$dF_Q = \sigma_Q dA = \frac{M_Q y}{I} dA \tag{8.20}$$

Since the P effect is a second force system

$$d\delta_P = \epsilon_P dx = \frac{\sigma_P}{E} dx = \frac{M_P y}{EI} dx \tag{8.21}$$

Thus,

$$IVW = \int_0^l \left[\int_A \frac{M_Q M_P}{EI^2} y^2 dA \right] dx = \int_0^l \frac{M_Q M_P}{EI} dx \tag{8.22}$$

As an example of the use of the method of virtual work to obtain displacements of a structure, consider the frame shown in the accompanying figure where the reactions are also shown.

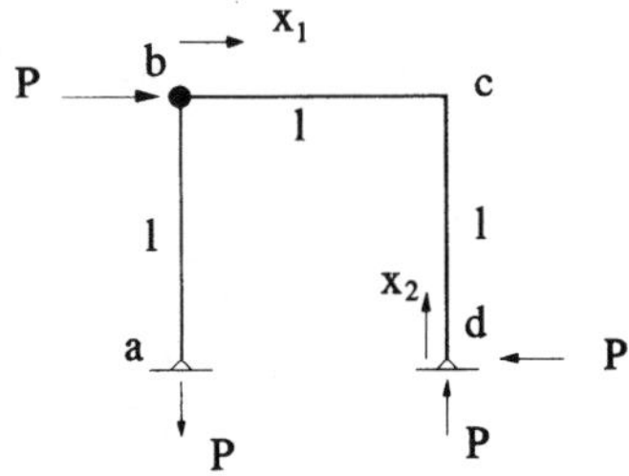

Let us find the horizontal displacement at point b. The first step is to select a Q-force system that will do work acting through the displacement we want to find. The next figure shows such a force system. Note that the computed reactions are part of the Q system since it must be an equilibrium force system.

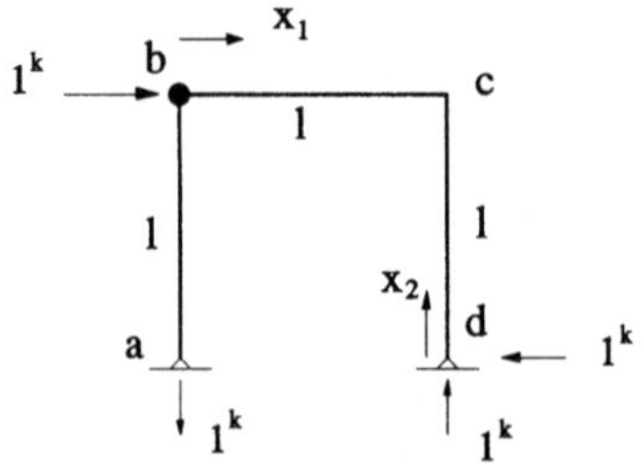

Q-force system.

The external virtual work, that is, the work done by the Q-force system acting through displacements caused by the P effect (the actual force system), is $(1^k)(\delta_b)$, where a positive δ_b acts in the direction of the 1^k load since the work is considered positive.

If we consider only bending effects, the internal virtual work is given by equation (8.22). This equation requires the bending-moment expressions for each member for each loading system. These expressions are summarized below.

Member *ab*:

$$M_P = 0 \qquad M_Q = 0$$

Member *bc*:

$$M_P = -Px_1 \qquad M_Q = -1\ x_1$$

Member *cd*:

$$M_P = -Px_2 \qquad M_Q = -1\ x_2$$

From equation (8.22), the internal virtual work becomes

$$IVW = \int_0^l \frac{Px_1^2}{EI} dx_1 + \int_0^l \frac{Px_2^2}{EI} dx_2 = \frac{2}{3} \frac{Pl^3}{EI}$$

Equating external and internal virtual work we have

$$(1^k)(\delta_b) = 2Pl^3/3EI$$

Thus,

$$\delta_b^{\text{horiz.}} = 2Pl^3/3EI$$

We next derive the elemental stiffness matrices for an axial force element and a beam element using the principle of virtual work.

8.3 ELEMENTAL STIFFNESS USING THE PRINCIPLE OF VIRTUAL WORK

8.3a Shape Functions and Elemental Stiffness for an Axially Loaded Bar

Consider the bar shown in Figure 8-12a, which is subjected to axial forces F_1 and F_2 at its ends and a distributed axial force $p(x)$ along its length. Figure 8-12b shows a small segment of the bar.

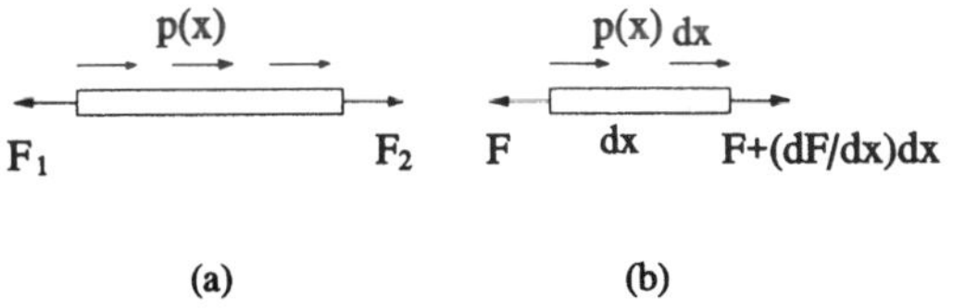

p(x)
F_1 F_2
(a)

p(x) dx
F dx F+(dF/dx)dx
(b)

Figure 8-12 Axial force member.

Equilibrium for the segment requires that

$$\frac{dF}{dx} + p(x) = 0 \tag{8.23}$$

Now, $F = \sigma A = EA\epsilon$. Also, from strength of materials, $\epsilon = du/dx$. Thus,

$$\frac{d}{dx}\left(EA\frac{du}{dx}\right) + p(x) = 0 \tag{8.24}$$

If there is no distributed load applied to the bar, then

$$\frac{d}{dx}\left(EA\frac{du}{dx}\right) = 0 \tag{8.25}$$

Assuming that the area of the bar, A, is constant, integration of equation (8.25) yields

$$EAu = C_1x + C_2 \tag{8.26}$$

We want to express the displacement of the bar in terms of the nodal displacements at its left and right ends, u_1 and u_2. With the origin of the coordinate system at the left end of the rod, $u(0) = u_1$ and $u(L) = u_2$. Applying these conditions to our displacement expression, equation (8.26), we find

$$EAu_1 = C_1 \text{ and } Eau_2 = C_1 L + C_2, \text{ from which } C_2 = (-EA/L)u_1 + (EA/L)u_2.$$

Substituting for C_1 and C_2 in our displacement equation, we find

$$u(x) = \left(1 - \frac{x}{L}\right) u_1 + \left(\frac{x}{L}\right) u_2 \tag{8.27}$$

The above equation gives the displacement u at any point along the length of the bar in terms of the nodal displacements u_1 and u_2. The expressions $1 - x/L$ and x/L are called *shape* or *basis* functions and will be notated as N_1 and N_2. Note that at the nodes, one shape function is unity and the other zero. These shape functions are plotted in Figure 8-13.

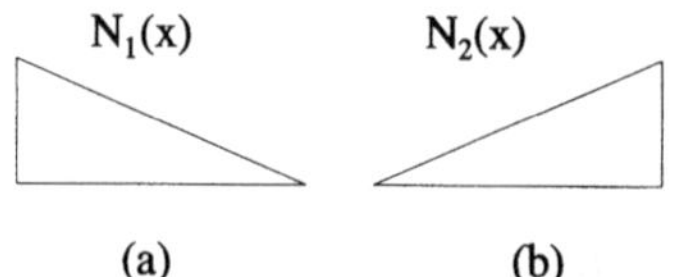

Figure 8-13 Shape functions for axial bar element.

In this case we were able to generate the shape functions by solving by exact means the differential equation governing the one-dimensional rod. For more complicated elements such as the triangular plate element, we will not be able to do this. The procedure we follow, which most often results in approximate shape functions, is to assume a displacement function containing as many unknown coefficients as we have degrees of freedom for the element. We then evaluate the coefficients in terms of the nodal displacements.

To illustrate for the one-dimensional rod, assume a displacement function containing two coefficients. Let us use a polynomial and assume

$$u(x) = ax + b \tag{8.28}$$

Now, at $x = 0$, $u(0) = u_1$ and at $x = L$, $u(L) = u_2$. Using these conditions in equation (8.28) we find

$$\begin{aligned} u(0) &= u_1 = b \\ u(L) &= u_2 = aL + b = aL + u_1 \end{aligned} \tag{8.29}$$

Thus, $b = u_1$, $a = (u_2 - u_1)/L$ and

$$u(x) = \left(\frac{u_2 - u_1}{L}\right) x + u_1 = \left(1 - \frac{x}{L}\right) u_1 + \frac{x}{L} u_2 \tag{8.30}$$

In this case we generated the same shape functions as we did by solving the differential equation. This occurred since the assumed polynomial was of the same form as the exact solution. Had we assumed a different displacement variation, say a sine function, the shape functions would have a different form. We now illustrate how the shape functions can be used in conjunction with the principle of virtual work for deformable bodies to derive the elemental stiffness matrix.

The expression for external virtual work is simply

$$\sum_{i=1}^{n} Q_n \delta^P$$

For axial stresses, if the P system was a second force system then

$$IVW = \int_0^L \left(\frac{F_P F_Q}{EA} \right) dx \tag{8.31}$$

Consider the displacement function when $u_1 = 1$ and $u_2 = 0$. In this case the displacement of the bar is $N_1(x)$. The forces required to maintain this displaced configuration are shown in Figure 8-14a. From the definition of stiffness, the forces at the ends of the member are actually elements of the stiffness matrix (forces corresponding to a unit displacement of one node and only one node). Similarly, Figure 8-14b illustrates the forces for the case where $u_1 = 0$ and $u_2 = 1$. The shape of the axial displacement curve is $N_2(x)$ in this case.

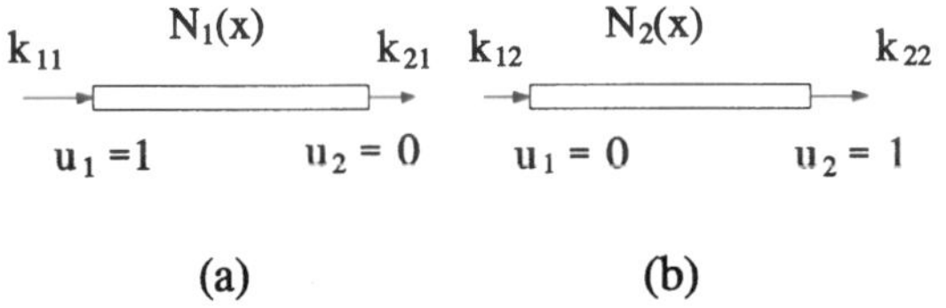

Figure 8-14

In applying the principle of virtual work, consider the bar of Figure 8-14a as the "P system" and that of Figure 8-14b as the "Q system." The external virtual work, that is the work done by the Q system acting through displacements caused by the P system, can be written:

External virtual work $= EVW = k_{12}u_1 + k_{22}u_2 = k_{12}(1) + k_{22}(0) = k_{12}$

Since $u(x) = N_1 u_1 + N_2 u_2$ and $F = EA\epsilon = EA\, du/dx$, then $F_P = EA\, dN_1/dx$ and $F_Q = EA\, dN_2/dx$. Therefore,

$$IVW = \int_0^L EA \left(\frac{dN_1}{dx} \right) \left(\frac{dN_2}{dx} \right) dx = EA \int_0^L \left(\frac{-1}{L} \right) \left(\frac{1}{L} \right) dx = -\frac{EA}{L} \tag{8.32}$$

Equating external and internal virtual work gives

$$k_{12} = -\frac{EA}{L} \tag{8.33}$$

Considering Figure 8-14a as both the P and Q systems we find

$$k_{11} = \int_0^L EA \left(\frac{dN_1}{dx} \right) \left(\frac{dN_1}{dx} \right) dx = EA \int_0^L \left(\frac{1}{L^2} \right) dx = \frac{EA}{L} \tag{8.34}$$

With Figure 8-14b as both P and Q systems we have

$$k_{22} = EA \int_0^L \left(\frac{1}{L^2}\right) dx = \frac{EA}{L} \tag{8.35}$$

Finally, using Figure 8-14b as the P system and Figure 8-14a as the Q system yields

$$k_{21} = k_{12} = -\frac{EA}{L} \tag{8.36}$$

In general, we can write

$$k_{ij} = \int_0^L EA \left(\frac{dN_i}{dx}\right)\left(\frac{dN_j}{dx}\right) dx \tag{8.37}$$

The final elemental stiffness matrix becomes

$$[K] = \frac{EA}{L}\begin{bmatrix} 1 & -1 \\ -1 & 1 \end{bmatrix} \tag{8.38}$$

Equation (8.38) is identical to that derived from strength of materials concepts.

8.3b Shape Functions and Elemental Stiffness for a Beam Element

Consider the beam element shown in Figure 8-15.

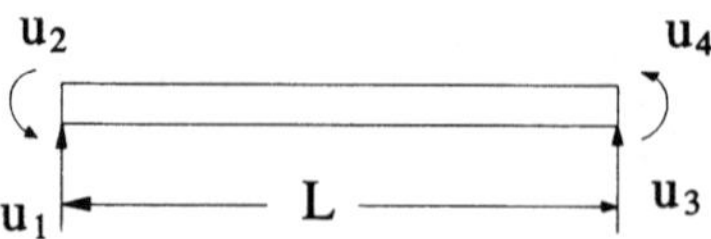

Figure 8-15 Beam element.

For a constant moment of inertia, the differential equation governing the displacement of the beam, assuming no applied distributed load, is

$$\frac{d^4y}{dx^4} = 0 \tag{8.39}$$

Integrating, we have

$$\frac{d^3y}{dx^3} = C_1 \tag{8.40}$$

$$\frac{d^2y}{dx^2} = C_1x + C_2 \tag{8.41}$$

$$\frac{dy}{dx} = C_1\left(\frac{x^2}{2}\right) + C_2x + C_3 \tag{8.42}$$

$$y(x) = C_1\left(\frac{x^3}{6}\right) + C_2\left(\frac{x^2}{2}\right) + C_3x + C_4 \tag{8.43}$$

Now,

$$\frac{dy}{dx}(0) = u_2 \qquad \frac{dy}{dx}(L) = u_4$$
$$y(0) = u_1 \qquad y(L) = u_3 \tag{8.44}$$

Applying the conditions specified in equation (8.44), solving for the constants C_1 through C_4, and back-substituting into equation (8.43), we find

$$y(x) = \left(\frac{2}{L^3}x^3 - \frac{3}{L^2}x^2 + 1\right)u_1 + \left(\frac{1}{L^2}x^3 - \frac{2}{L}x^2 + x\right)u_2$$
$$+ \left(\frac{3}{L^2}x^2 - \frac{2}{L^3}x^3\right)u_3 + \left(\frac{1}{L^2}x^3 - \frac{1}{L}x^2\right)u_4 \tag{8.45}$$

or

$$y(x) = N_1u_1 + N_2u_2 + N_3u_3 + N_4u_4 \tag{8.46}$$

These shape functions are plotted in Figure 8-16.

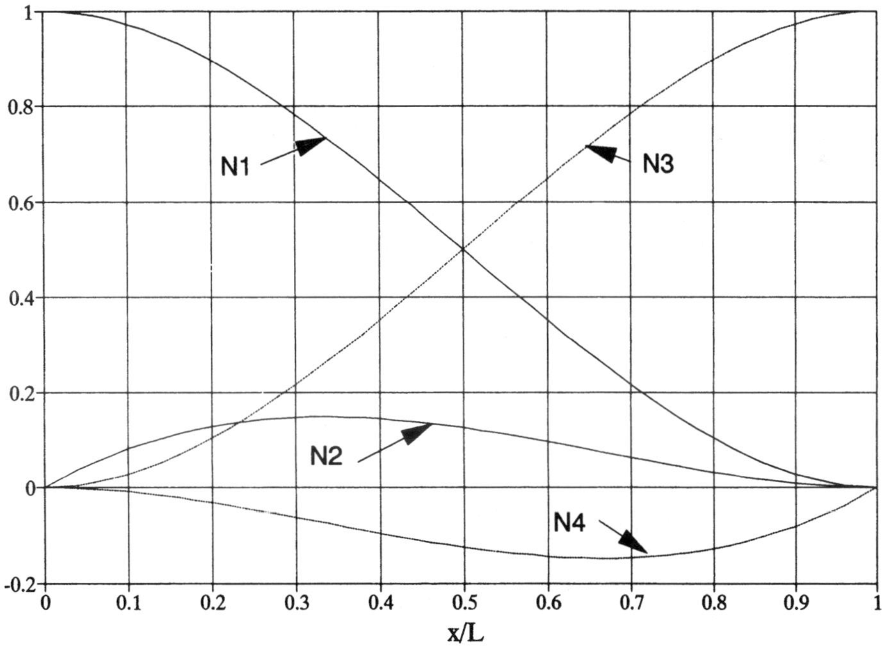

Figure 8-16 Shape functions for a beam element.

Next we consider the determination of k_{11} by using virtual work and the shape functions. Figure 8-17 will be used for both the Q and P systems.

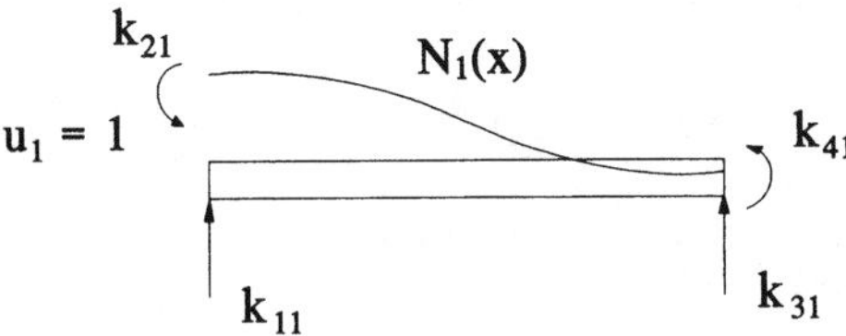

Figure 8-17 *P* and *Q* systems for k_{11}.

The moment-curvature relationship is

$$M = EI\frac{d^2y}{dx^2} \tag{8.47}$$

Thus,

$$M_P = M_Q = EI\frac{d^2N_1}{dx^2} \tag{8.48}$$

and

$$k_{11}(1) = \int_0^L EI\left(\frac{d^2N_1}{dx^2}\right)\left(\frac{d^2N_1}{dx^2}\right)dx \tag{8.49}$$

Integrating equation (8.49) we find $k_{11} = 12EI/L^3$.

Consider the stiffness coefficient k_{21}. We shall use the shape functions N_1 and N_2 as shown in Figure 8-18.

$$EVW = k_{21}u_2 = k_{21} = \int_0^L EI\left(\frac{d^2N_1}{dx^2}\right)\left(\frac{d^2N_2}{dx^2}\right)dx \tag{8.50}$$

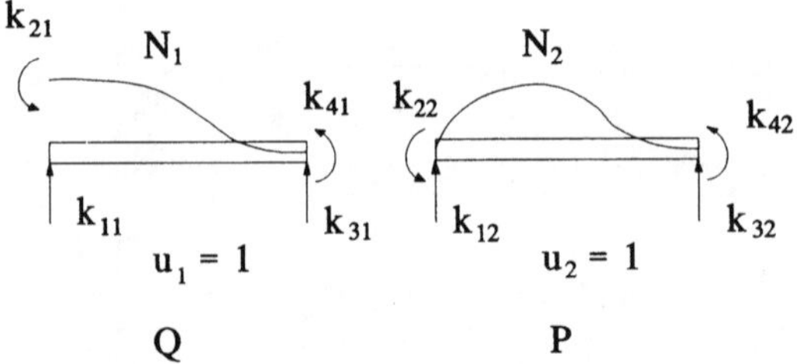

Figure 8-18 *P* and *Q* systems for k_{21}.

Performing the indicated operations in equation (8.50) we find $k_{21} = 6EI/L^2$.

In general,

$$k_{ij} = \int_0^L EI\left(\frac{d^2N_i}{dx^2}\right)\left(\frac{d^2N_j}{dx^2}\right)dx \tag{8.51}$$

Of course, we could assume a displacement function as we did in the case of the axial force member. If we assume

$$y(x) = ax^3 + bx^2 + cx + d \tag{8.52}$$

and apply the conditions indicated by equation (8.42), we will generate the same shape functions as those in equation (8.43). This occurs because the form of our assumed displacement variation is identical to the form generated by the solution to the differential equation.

8.3c Shape Functions and Elemental Stiffness for a Three-Node Axial Element

We can develop elemental stiffnesses for many different elements by the use of shape functions. In fact, this is exactly what is done in finite element analysis. As an example, consider the three-node bar element shown in Figure 8-19.

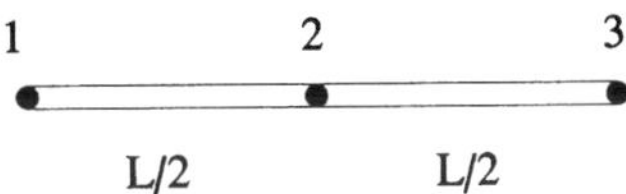

Figure 8-19 Three-node bar element.

We assume the form for the displacement function shown in equation (8.53).

$$u(x) = a_o + a_1 x + a_2 x^2 \tag{8.53}$$

Equations (8.54) indicate the conditions we impose in order to solve for the three unknown coefficients in terms of the nodal displacements.

$$\begin{aligned} u(0) &= u_1 = a_0 \\ u(L) &= u_3 = u_1 + a_1 L + a_2 L^2 \\ u(L/2) &= u_2 = u_1 + \frac{a_1 L}{2} + \frac{a_2 L^2}{4} \end{aligned} \tag{8.54}$$

Solving (8.54) for the constants a_0, a_1, and a_2, and then substituting the results into equation (8.53), we find

$$u(x) = u_1 \left[1 - \frac{3x}{L} + \frac{2x^2}{L^2}\right] + u_2 \left[\frac{4x}{L} - \frac{4x^2}{L^2}\right] + u_3 \left[\frac{2x^2}{L^2} - \frac{x}{L}\right] \tag{8.55}$$

The coefficients of the u_i displacements are the shape functions N_1, N_2, and N_3.

Since we are dealing with an axial force member, equation (8.37) for the k_{ij}'s is still valid. The i and j subscripts will take on values of 1, 2, and 3 in this case. Performing the operations required by equation (8.37) we find for the stiffness matrix for the three node bar element

$$[k] = \frac{EA}{3L} \begin{bmatrix} 7 & -8 & 1 \\ -8 & 16 & -8 \\ 1 & -8 & 7 \end{bmatrix} \tag{8.56}$$

8.4 NON-NODAL FORCES USING THE CONCEPT OF EQUIVALENT WORK

When we dealt with non-nodal forces earlier in the text we found "fixed end" forces corresponding to the applied loads and then reversed their direction. These forces were used as nodal loads and resulted in the true nodal displacements. Of course, after finding the member forces using the calculated displacements, we applied the fixed end forces to the nodal points to determine the actual member forces at the nodes. We shall now use shape functions and the concept of equivalent work to determine equivalent nodal forces.

We will define an equivalent force as one that does work equal to the work done by the original force when acting through the same imposed displacement. We shall use the shape functions as a convenient set of displacements through which these forces act.

Figure 8-20 shows an axial force member loaded with a distributed load $p(x)$. Also shown are equivalent nodal forces, F_{1e} and F_{2e}, and the two shape functions, $N_1(x)$ and $N_2(x)$.

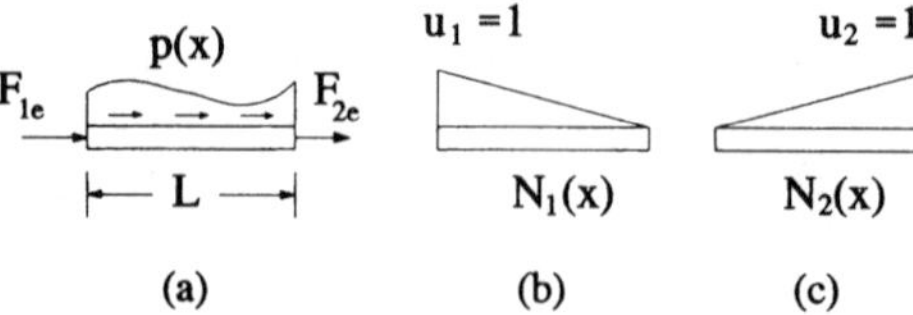

Figure 8-20 Non-nodal load, equivalent nodal forces (**a**), and shape functions (**b,c**).

The work done by the equivalent nodal forces in Figure 8-20a acting through the displacements of Figure 8-20b, that is, $N_1(x)$, is simply $F_{1e}u_1 = F_{1e}$.

The actual force acting on a dx length of the bar is $p(x)dx$. The work done by this force is $p(x)dxN_1(x)$. The total work is therefore the integral of this quantity taken over the loaded length of the bar; in this example, the total length of the element L.

Equating the work done by the equivalent nodal forces to that of the applied load $p(x)$, we have

$$F_{1e} = \int_0^L p(x)N_1(x)dx \tag{8.57}$$

Using Figure 8-20c as the displacement we find

$$F_{2e} = \int_0^L p(x)N_2(x)dx \tag{8.58}$$

In general,

$$F_{ie} = \int_0^L p(x)N_i(x)dx \tag{8.59}$$

To illustrate, assume that $p(x)$ is uniform over the length of the bar. That is, $p(x) = p_0$. Then,

$$F_{1e} = \int_0^L p_0 N_1(x)dx = p_0 \int_0^L \left(1 - \frac{x}{L}\right) dx = \frac{p_0 L}{2} \tag{8.60}$$

Also,

$$F_{2e} = \int_0^L p_0 N_2(x)dx = p_0 \int_0^L \left(\frac{x}{L}\right) dx = \frac{p_0 L}{2} \tag{8.61}$$

As you would expect intuitively, half the load acts at each node.

As a second example, consider the triangular load shown in Figure 8-21.

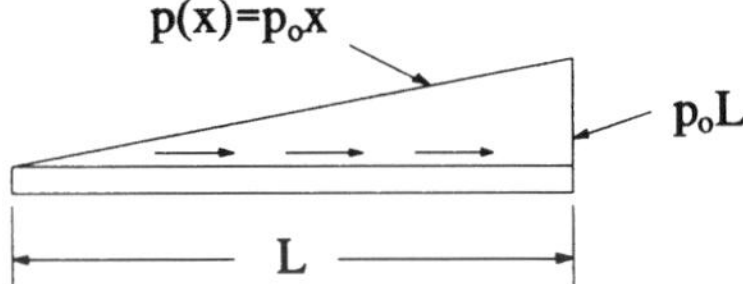

Figure 8-21

Equation (8.59) yields

$$F_{1e} = \int_0^L p_0 x N_1(x)dx = p_0 \int_0^L x\left(1 - \frac{x}{L}\right) dx = \frac{p_0 L^2}{6} \tag{8.62}$$

and

$$F_{2e} = \int_0^L p_0 x N_2(x)dx = p_0 \int_0^L x\left(\frac{x}{L}\right) dx = \frac{p_0 L^2}{3} \tag{8.63}$$

Since the total load applied to the bar is $p_0 L^2/2$, we see that the equivalent forces F_{1e} and F_{2e} are one-third and two-thirds of the total load, respectively.

As another example consider the beam shown in Figure 8-22, which is subjected to a positive (upward) uniformly distributed load.

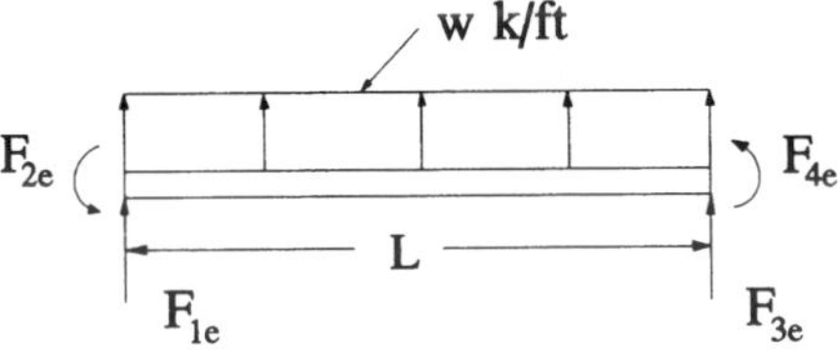

Figure 8-22

Using the shape functions in equation (8.45) we find

$$F_{1e} = w \int_0^L \left(\frac{2x^3}{L^3} - \frac{3x^2}{L^2} + 1\right) dx = \frac{wL}{2} \tag{8.64}$$

$$F_{3e} = w \int_0^L \left(\frac{3x^2}{L^2} - \frac{2x^3}{L^3}\right) dx = \frac{wL}{2} \tag{8.65}$$

$$F_{2e} = w \int_0^L \left(\frac{x^3}{L^2} - \frac{2x^2}{L} + x \right) dx = \frac{wL^2}{12} \tag{8.66}$$

$$F_{4e} = w \int_0^L \left(\frac{x^3}{L^2} - \frac{x^2}{L} \right) dx = -\frac{wL^2}{12} \tag{8.67}$$

Next consider the case of a beam with a concentrated force applied as shown in Figure 8-23. Note that the load is acting down and is therefore negative.

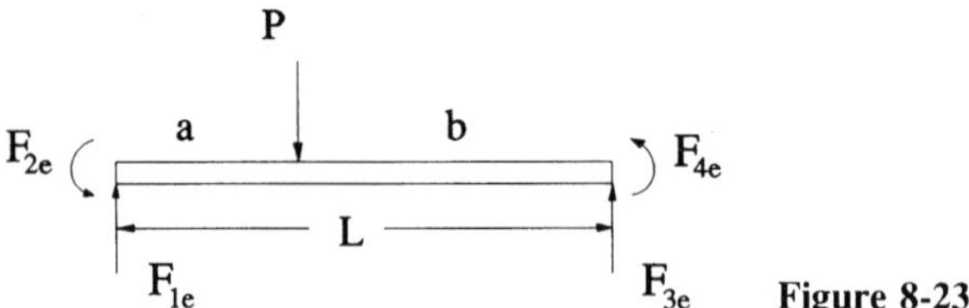

Figure 8-23

In this case, the integral in equation (8.59) becomes an evaluation at the point of application of the force P.

$$F_{1e} = -PN_1(a) = -P\left(\frac{2a^3}{L^3} - \frac{3a^2}{L^2} + 1 \right) = -\frac{Pb^2}{L^3}(3a + b) \tag{8.68}$$

$$F_{2e} = -PN_2(a) = -P\left(\frac{a^3}{L^2} - \frac{2a^2}{L} + a \right) = -\frac{Pab^2}{L^2} \tag{8.69}$$

The other equivalent nodal forces are found in a similar manner. Note that equations (8.68) and (8.69) are identical in magnitude to the fixed end forces and moments that were presented in Chapter 3 for this loading condition.

8.5 STRAIN ENERGY AND FORCE POTENTIAL

8.5a Strain Energy

Consider the axially loaded bar shown in Figure 8-24.

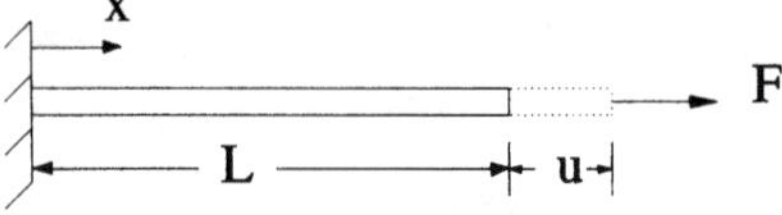

Figure 8-24 Axially loaded bar.

As the force is slowly increased to its full value (static loading), the bar continues to change in length. Since we are dealing with a linear system, the force-displacement relationship for the bar is as shown in Figure 8-25.

The work done by the force acting through the displacement it causes is equal to the area under the force-displacement curve. For the linear system shown, the work equals $(1/2)Fu$.

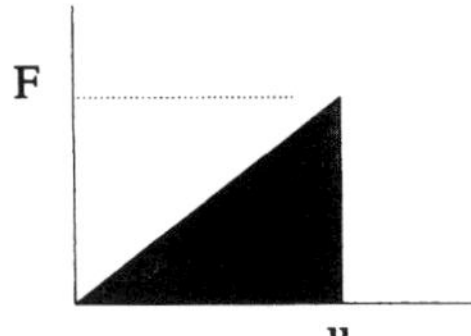

Figure 8-25 Force-displacement relationship.

Since work has been performed on the bar, its internal energy must increase. Assuming no energy losses such as heat dissipation, this increase in internal energy must be equal to the work done by the force. This internal energy is called *strain energy* since it is due to deformation of the bar, and is fully recoverable. Thus, it is potential energy stored in the bar. We generally call it strain energy, however, to distinguish it from the potential energy of other forces acting on the body.

We can rewrite the strain energy in terms of strain energy per unit volume by using relationships between force and stress, and between displacements and strain. For the bar shown in Figure 8-24, the force is constant throughout the member. Assuming the area is also constant, then $F = \sigma A$ and $u = \epsilon L$. Thus, $(1/2)Fu = (1/2)\sigma A \epsilon L$.

The volume of the bar is $A \times L$. Thus the strain energy per unit volume is $(1/2)\sigma\epsilon$. For an infinitesimal element with volume dV, the strain energy U stored is given by

$$dU = \frac{1}{2}\sigma\epsilon dV \tag{8.70}$$

For this one-dimensional state of stress in the x direction, this expression is generally written

$$dU = \frac{1}{2}\sigma_x\epsilon_x dV \tag{8.71}$$

Naturally, the expression for strain energy can be extended to two- and three-dimensional states of stress. For a general two-dimensional state of stress we find

$$dU = \left[\frac{1}{2}\sigma_x\epsilon_x + \frac{1}{2}\sigma_y\epsilon_y + \frac{1}{2}\tau_{xy}\gamma_{xy}\right] dV \tag{8.72}$$

The total strain energy is found by integrating equation (8.72) over the volume of the body. Considering the two-dimensional case we have

$$U = \int_V \frac{1}{2}\{\epsilon\}^T\{\sigma\} dV \tag{8.73}$$

where

$$\{\epsilon\} = \begin{Bmatrix} \epsilon_x \\ \epsilon_y \\ \gamma_{xy} \end{Bmatrix} \quad \text{and} \quad \{\sigma\} = \begin{Bmatrix} \sigma_x \\ \sigma_y \\ \tau_{xy} \end{Bmatrix} \tag{8.74}$$

Note that ϵ^T is used in equation (8.73) to make the matrix multiplication conformable. That is,

$$\{\epsilon\}^T = \left[\epsilon_x\, \epsilon_y\, \gamma_{xy}\right] \tag{8.75}$$

and

$$\{\epsilon\}^T \{\sigma\} = \epsilon_x\sigma_x + \epsilon_y\sigma_y + \gamma_{xy}\tau_{xy} \tag{8.76}$$

For the one-dimensional element,

$$\sigma_x = E\epsilon_x \quad \text{and} \quad U = \int_0^L \frac{EA}{2}\epsilon_x^T \epsilon_x dx = \int_0^L \frac{EA}{2}\epsilon_x^2 dx \tag{8.77}$$

since ϵ_x is the only non-zero strain.

8.5b Force Potential

The work done by a force vector $\vec{\mathbf{F}}$ acting through a displacement is given by

$$W = \int_1^2 \vec{\mathbf{F}} \cdot d\vec{r} \tag{8.78}$$

where points 1 and 2 are the beginning and ending points on the displacement curve (see Figure 8-26).

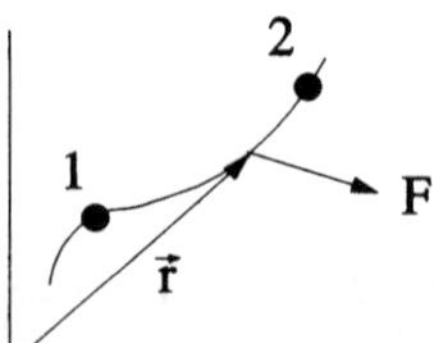

Figure 8-26

In general, the work depends on the path of integration as well as the end points. However, there are forces for which the work integral does not depend on the path of integration. These forces are called *conservative forces*, and the work done can be determined by the difference of a scalar function evaluated at the end points of the integration path. Also, if the magnitude of the force is not dependent on its displacement, the potential energy is the negative of the dot product between the force and displacement vectors—that is, the opposite of the work done by the force as it is moved through the displacement. Thus, we have

$$V = -F_x u - F_y v - F_z w \tag{8.79}$$

For nodal forces of constant magnitude F_i, the nodal force potential at node i for the one-dimensional rod becomes $V_i = -F_i u_i$ and at node j is $V_j = -F_j u_j$. Thus, for nodal forces acting on the bar element,

$$V_{nodal\ force} = -\left[\, F_i \quad F_j \,\right] \begin{Bmatrix} u_i \\ u_j \end{Bmatrix} \tag{8.80}$$

Next, consider a distributed force applied to the element as shown in Figure 8-27.

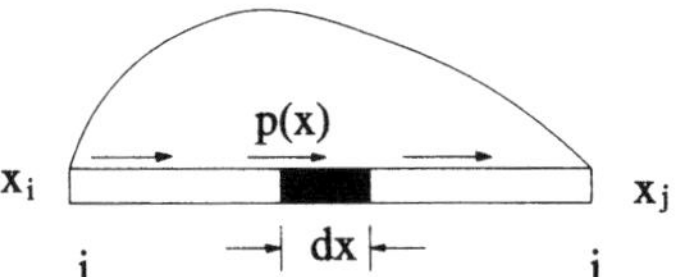

Figure 8-27

For element dx, $dF = p(x)dx$ and $dV_{\text{dist.}} = -p(x)u(x)dx$. Remembering that

$$u = \left[1 - \frac{x}{L} \quad \frac{x}{L}\right] \begin{Bmatrix} u_i \\ u_j \end{Bmatrix} \tag{8.81}$$

then the total potential energy of the distributed load becomes

$$V_{dist.} = -\int_{x_i}^{x_j} p(x) \left[1 - \frac{x}{L} \quad \frac{x}{L}\right] \begin{Bmatrix} u_i \\ u_j \end{Bmatrix} dx = -\int_{x_i}^{x_j} p(x) \left[\, N_1 \quad N_2 \,\right] \begin{Bmatrix} u_i \\ u_j \end{Bmatrix} dx \tag{8.82}$$

As an example, consider the bar loaded as shown in Figure 8-28.

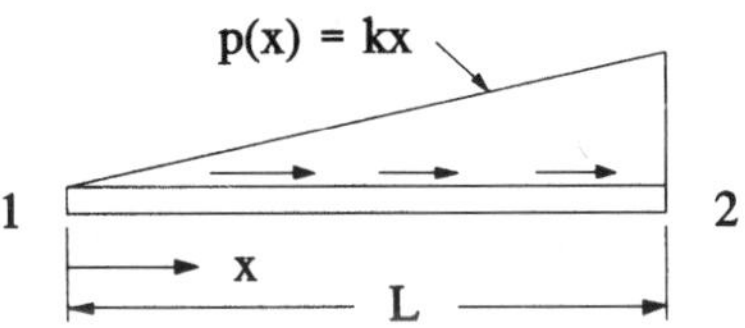

Figure 8-28 Linearly distributed bar loading.

Equation (8.82) becomes

$$V_{dist.} = -\int_0^L kx \left[1 - \frac{x}{L} \quad \frac{x}{L}\right] \begin{Bmatrix} u_1 \\ u_2 \end{Bmatrix} dx = -\frac{kL^2}{2} \begin{bmatrix} \frac{1}{3} & \frac{2}{3} \end{bmatrix} \begin{Bmatrix} u_1 \\ u_2 \end{Bmatrix} \tag{8.83}$$

Note that the total load applied to the bar is $kL^2/2$. Since the potential energy is the negative of the work done, then the equivalent nodal loads are $F_1 = 1/3 \times$ (total load) and $F_2 = 2/3 \times$ (total load). Since

$$U = \int_0^L \frac{EA}{2} \epsilon_x^T \epsilon_x dx \tag{8.84}$$

and $\epsilon_x = du/dx$, and using

$$u = \begin{bmatrix} 1 - \frac{x}{L} & \frac{x}{L} \end{bmatrix} \begin{Bmatrix} u_i \\ u_j \end{Bmatrix} \tag{8.85}$$

we have the strain-displacement relationship

$$\epsilon_x = \frac{du}{dx} = \begin{bmatrix} -\frac{1}{L} & \frac{1}{L} \end{bmatrix} \begin{Bmatrix} u_i \\ u_j \end{Bmatrix} = \frac{1}{L} [\, -1 \quad 1 \,] \begin{Bmatrix} u_i \\ u_j \end{Bmatrix} \tag{8.86}$$

The strain energy becomes

$$\begin{aligned} U &= \int_0^L \frac{EA}{2} \epsilon_x^T \epsilon_x dx = \int_0^L \frac{EA}{2} \frac{1}{L} [\, u_i \quad u_j \,] \begin{Bmatrix} -1 \\ 1 \end{Bmatrix} \frac{1}{L} [\, -1 \quad 1 \,] \begin{Bmatrix} u_i \\ u_j \end{Bmatrix} dx \\ &= \int_0^L \frac{EA}{2L^2} [\, u_i \quad u_j \,] \begin{bmatrix} 1 & -1 \\ -1 & 1 \end{bmatrix} \begin{Bmatrix} u_i \\ u_j \end{Bmatrix} dx \\ &= \frac{EA}{2L} [\, u_i \quad u_j \,] \begin{bmatrix} 1 & -1 \\ -1 & 1 \end{bmatrix} \begin{Bmatrix} u_i \\ u_j \end{Bmatrix} = \frac{1}{2} \{u\}^T [K] \{u\} \end{aligned} \tag{8.87}$$

where

$$[K] = \frac{EA}{L} \begin{bmatrix} 1 & -1 \\ -1 & 1 \end{bmatrix} \tag{8.88}$$

Note that we could have obtained this general form using $U = (1/2)Fu$. In matrix form $U = 1/2\{F\}^T\{u\} = 1/2\{F\}\{u\}^T$. Also, $\{F\} = [K]\{u\}$ and $\{F\}^T = \{u\}^T[K]^T = \{u\}^T[K]$ since $[K]$ is symmetric. Thus,

$$U = \frac{1}{2} \{u\}^T [K] \{u\} \tag{8.89}$$

8.6 THE PRINCIPLE OF MINIMUM POTENTIAL ENERGY

The total potential energy Π is equal to the sum of the strain energy, nodal force potential, distributed force potential, and body force potential. Body forces are forces per unit volume acting on a body. Gravity is an example of a body force. Considering only the nodal and distributed surface forces we have

$$\begin{aligned} \Pi &= U + V_{nodal} + V_{dist.} \\ &= \frac{1}{2} [\, u_i \quad u_j \,] [K] \begin{Bmatrix} u_i \\ u_j \end{Bmatrix} - [\, F_i \quad F_j \,] \begin{Bmatrix} u_i \\ u_j \end{Bmatrix} \\ &\quad - \int_{x_i}^{x_j} p(x) [\, N_1 \quad N_2 \,] \begin{Bmatrix} u_i \\ u_j \end{Bmatrix} dx \end{aligned} \tag{8.90}$$

where N_1 and N_2 are the shape functions.

We now state the principle of minimum potential energy.

Of all possible displacement fields which satisfy the specified geometric constraints, the true one, which corresponds to a stable equilibrium state, minimizes the potential energy Π.

This principle can be used to obtain approximate solutions to many types of problems and can also be used to derive the force-displacement relationships for a single element or a system containing many elements. We first illustrate its use in approximate analysis.

8.7 APPROXIMATE SOLUTIONS USING MINIMUM POTENTIAL ENERGY

We now outline the procedure for solving problems approximately, using the principle of minimum potential energy.

(1) Select a displacement function that satisfies the geometric boundary conditions. This function must contain some arbitrary constants that will be determined after application of the minimization process.
(2) Determine the expression for the potential energy (Π) using the assumed displacement function.
(3) Minimize the potential energy by differentiating with respect to each constant and solve the resulting equations for these constants.

Example 8.1

Suppose we would like to find a solution for the displacement as a function of position for the tapered bar shown in Figure 8-29.

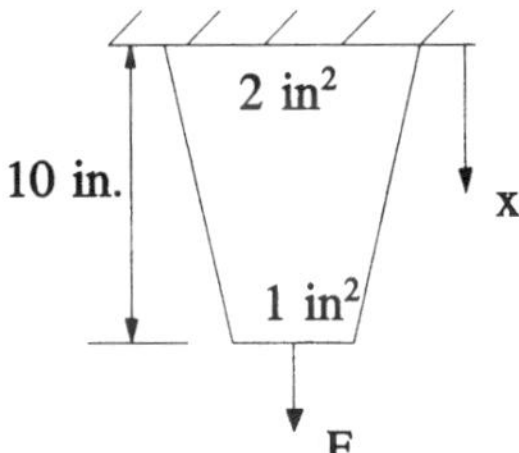

Figure 8-29

Of course, for this problem the exact solution can be found as follows:

$$A(x) = 2 - \frac{x}{10}$$

$$\epsilon(x) = \frac{F}{EA(x)} = \frac{F/E}{\left(2 - \frac{x}{10}\right)} \tag{8.91}$$

$$u(x) = \int \epsilon(x)dx = \int \frac{F/E}{\left(2 - \frac{x}{10}\right)}dx = -10\frac{F}{E}\ln\left(2 - \frac{x}{10}\right) + C_1 \tag{8.92}$$

Applying the condition $u(0) = 0$ yields $C_1 = 10(F/E)\ln 2$. Thus,

$$u_{exact}(x) = \frac{10F}{E}\left[\ln 2 - \ln\left(2 - \frac{x}{10}\right)\right] \tag{8.93}$$

Now let us find an approximate solution to this problem by using the principle of minimum potential energy.

Step 1: Assume a displacement function that satisfies the geometric boundary conditions.

Assume

$$u(x) = a + bx + cx^2 \tag{8.94}$$

For

$$u(0) = 0, a = 0. \qquad \text{Thus } u(x) = bx + cx^2.$$

Step 2: Determine the total potential energy using the assumed displacement function.

The strain is given by equation (8.95), and the strain energy by equation (8.96).

$$\epsilon(x) = \frac{du}{dx} = b + 2cx \tag{8.95}$$

$$U = \frac{E}{2}\int_0^L \epsilon_x^2 A(x)dx = \frac{E}{2}\left(2b^2L + 4bcL^2 + \frac{8c^2L^3}{3} - \frac{b^2L^2}{20} - \frac{4bcL^3}{30} - \frac{4c^2L^4}{40}\right) \tag{8.96}$$

The force potential $V = -Fu(10) = -F(10b + 100c)$

The total potential energy is $\mathbf{\Pi} = U + V$.

Step 3: Differentiate the potential energy with respect to each of the arbitrary constants, setting the resulting equations equal to zero to minimize, and solve for the values of the constants.

$$\begin{aligned} \frac{\partial \Pi}{\partial b} &= 0 = \frac{E}{2}(30b + 266.66c) - 10F = 0 \\ \frac{\partial \Pi}{\partial b} &= 0 = \frac{E}{2}(266.66b + 3333.33c) - 100F = 0 \end{aligned} \tag{8.97}$$

Solving equations (8.97) for the values of b and c yields $b = 0.461541\ F/E$ and $c = 0.023077\ F/E$. Thus,

$$u_{approx.} = \frac{F}{E}\left(0.461541x + 0.023077x^2\right) \tag{8.98}$$

A comparison of the exact and approximate solutions for displacements is shown in Figure 8-30.

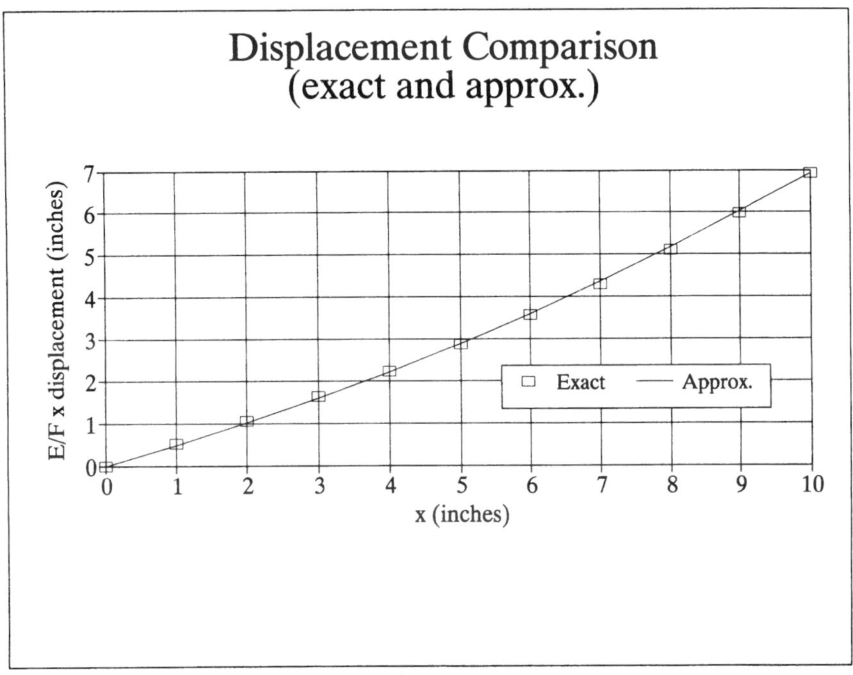

Figure 8-30

The axial strains $\epsilon_x = du/dx$, thus

$$\begin{aligned} \epsilon_{exact}(x) &= \frac{F/E}{\left(2 - \frac{x}{10}\right)} \\ \epsilon_{approx.} &= \frac{F}{E}(0.461541 + 0.046154x) \end{aligned} \tag{8.99}$$

Comparison of the strains is shown in Figure 8-31.

As can be seen from the comparison graphs, the exact and approximate displacements are very close to each other. The maximum displacements differ by less than 0.2%. However, the strains do not compare as well, the maximum difference being about 8%. This is to be expected given the form of the displacement function. The approximate expression for strain is a linear function of x and cannot accurately represent the exact strain variation. If more terms of the polynomial in x were added to the assumed displacement function, a better approximation of strain would result.

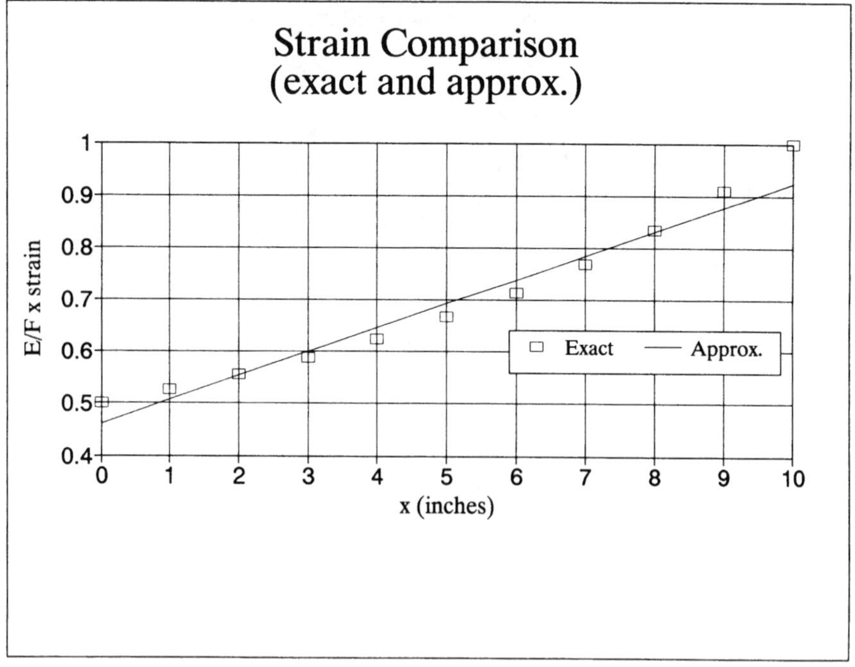

Figure 8-31

8.8 DETERMINATION OF THE STRUCTURAL STIFFNESS EQUATION USING MINIMUM POTENTIAL ENERGY

The total potential energy of a structure is simply the sum of potential energies of the individual elements of that structure. Consider the two-element axial structure shown in Figure 8-32.

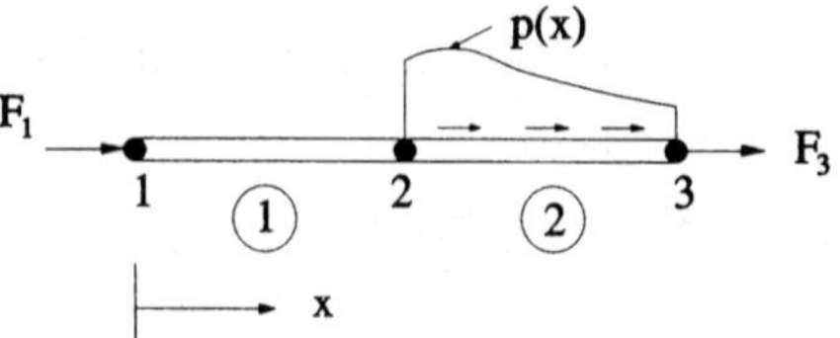

Figure 8-32 Two-element axial structure.

The potential energy of the structure can be written

$$\Pi = \frac{1}{2}[u_1 \quad u_2][k_1]\begin{Bmatrix} u_1 \\ u_2 \end{Bmatrix} + \frac{1}{2}[u_2 \quad u_3][k_2]\begin{Bmatrix} u_2 \\ u_3 \end{Bmatrix}$$

$$- F_1u_1 - F_3u_3 - \int_{x_2}^{x_3} p(x)[N_1 \quad N_2]\begin{Bmatrix} u_2 \\ u_3 \end{Bmatrix} dx \tag{8.100}$$

Defining the global displacement matrix {u} as

$$\{u\} = \begin{Bmatrix} u_1 \\ u_2 \\ u_3 \end{Bmatrix} \tag{8.101}$$

equation (8.100) can be written

$$\begin{aligned} \Pi = &- [\,F_1 \quad 0 \quad F_3\,]\{u\} + \frac{1}{2}\{u\}^T \begin{bmatrix} k_1 & \\ & k_2 \end{bmatrix} \{u\} \\ &- [\,0 \quad \textstyle\int_{x_2}^{x_3} p(x)N_1 dx \quad \int_{x_2}^{x_3} p(x)N_2 dx\,]\{u\} \end{aligned} \tag{8.102}$$

We now minimize the potential energy with respect to each of the unknown nodal displacements. For example:

$$\begin{aligned} \frac{\partial \Pi}{\partial u_1} = &- [\,F_1 \quad 0 \quad F_3\,]\begin{Bmatrix} 1 \\ 0 \\ 0 \end{Bmatrix} + \frac{1}{2}[\,1 \quad 0 \quad 0\,]\begin{bmatrix} & & \\ & K & \\ & & \end{bmatrix}\begin{Bmatrix} u_1 \\ u_2 \\ u_3 \end{Bmatrix} \\ &+ \frac{1}{2}[\,u_1 \quad u_2 \quad u_3\,]\begin{bmatrix} & & \\ & K & \\ & & \end{bmatrix}\begin{Bmatrix} 1 \\ 0 \\ 0 \end{Bmatrix} \\ &- [\,0 \quad \textstyle\int_{x_2}^{x_3} p(x)N_1 dx \quad \int_{x_2}^{x_3} p(x)N_2 dx\,]\begin{Bmatrix} 1 \\ 0 \\ 0 \end{Bmatrix} = 0 \end{aligned} \tag{8.103}$$

where

$$[K] = \begin{bmatrix} k_{11}^{(1)} & k_{12}^{(1)} & 0 \\ k_{12}^{(1)} & k_{22}^{(1)} + k_{11}^{(2)} & k_{12}^{(2)} \\ 0 & k_{21}^{(2)} & k_{22}^{(2)} \end{bmatrix} \tag{8.104}$$

Expanding and collecting terms we find

$$-F_1 + k_{11}^{(1)}u_1 + k_{12}^{(1)}u_2 - 0 = 0 \tag{8.105}$$

Differentiating the potential energy with respect to u_2 and then with respect to u_3 yields equations (8.106) and (8.107).

$$0 + k_{12}^{(1)}u_1 + \left(k_{22}^{(1)} + k_{11}^{(2)}\right)u_2 + k_{12}^{(2)}u_3 - \int_{x_2}^{x_3} p(x)N_1 dx = 0 \tag{8.106}$$

$$-F_3 + k_{21}^{(2)}u_2 + k_{22}^{(2)}u_3 - \int_{x_2}^{x_3} p(x)N_2 dx = 0 \tag{8.107}$$

Writing equations (8.105), (8.106), and (8.107) in matrix form gives

$$[K]\{u\} = \{F\}_{nodal} + \{F\}_{equiv.\ nodal} \tag{8.108}$$

Of course, this equation is identical to that obtained previously by direct combination.

8.9 SUMMARY

In this chapter we introduced important work and energy concepts and demonstrated how their use results in the determination of equivalent nodal forces and the structural stiffness equation, both of which were derived by direct means in earlier chapters. The methods presented represent an alternative approach for deriving the quantities needed and the equations used in matrix analysis of structures. Energy methods and minimization principles are used extensively in the development of the finite element method. This brief introduction to the use of these methods should begin to prepare you for a more advanced course in the finite element method.

PROBLEMS

8.1 Determine the shape functions for a one-dimensional rod by assuming the following expressions for the displacement field:

(a) $u(x) = \text{a} \ \sin \ x + b$

(b) $u(x) = \text{a} \ \sin(\pi x/2L) + b$

8.2 Determine the remaining elements of the stiffness matrix for a beam using the shape functions of equation (8.45) and the principle of virtual work.

8.3 Verify the stiffness terms in equation (8.56) for the three-node bar element.

For the following problems, verify the equivalent nodal forces using the methods of section 8.4.

8.4 Problem 1-5a

8.5 Problem 1-5c

8.6 Problem 1-5d

8.7 Problem 1-5e

8.8 Problem 1-5f

8.9 Find the equivalent nodal forces for the beam loaded as shown in Figure P8-9.

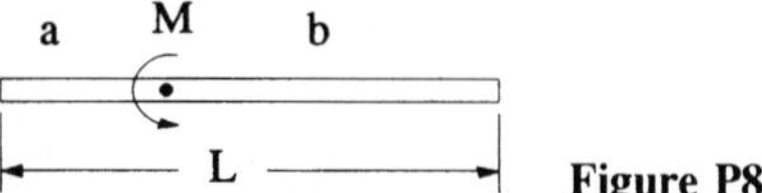

Figure P8-9

8.10 Show that the strain energy U for a beam is given by

$$U = \int_0^L \frac{EI}{2} \left(\frac{d^2 y}{dx^2} \right)^2 dx$$

Consider only bending effects.

8.11 Find the exact solution for the displacements and strains for the tapered bar shown in Figure P8-11.

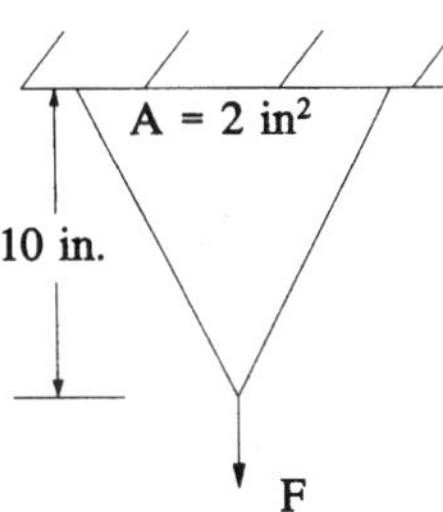

Figure P8-11

8.12 Find an approximate solution to problem 8.11 by using the principle of minimum potential energy. Graphically compare your results with the exact solution.

8.13 Use the expression for strain energy for a beam as given in problem 8.10 to find an approximate solution for the deflection of a simply supported beam with a concentrated load applied at the midpoint. Use minimum potential energy and compare your result for deflection at the midpoint with the exact solution of $PL^3/48EI$. Assume $y = a \sin(\pi x/L)$ for the displacement function.

8.14 Find an approximate solution for the deflection at the end of a cantilever beam with a concentrated load P applied at the end. Assume a fourth-order polynomial for the displacement function. Compare with the exact solution of $PL^3/3EI$.

8.15 Verify equations (8.106) and (8.107).

CHAPTER 9

A BRIEF INTRODUCTION TO THE FINITE ELEMENT METHOD

9.1 INTRODUCTION

In Chapter 8 we introduced the concepts of shape functions, virtual work, and minimum potential energy, and we used these concepts to derive the elemental stiffness matrix, equivalent nodal forces, and the structural stiffness equation. Of course, the results obtained were identical to those found in earlier chapters by direct means.

It was also noted that for more complicated elements, use of the techniques presented were necessary to develop the elemental stiffness matrix and the structural stiffness equation. The simplest element requiring the use of these techniques is the three-node triangular element subjected to loads in its plane. Figure 9-1 shows such an element.

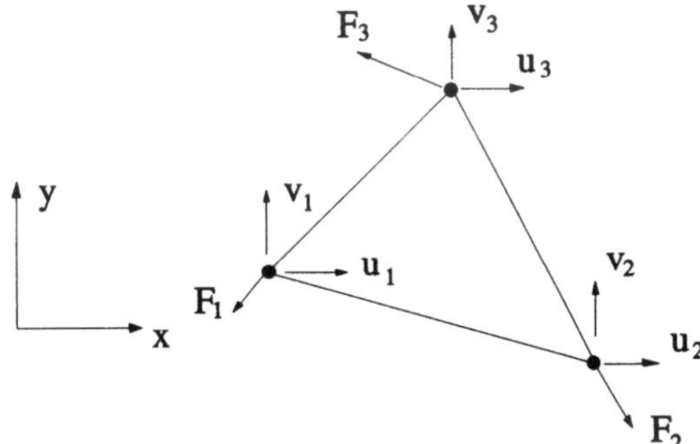

Figure 9-1 Three-node triangular element.

Also shown in Figure 9-1 are the nodal displacements u_1 through u_3 and v_1 through v_3, where the u and v displacements are in the global x and y directions, respectively.

Figure 9-2 shows how this element could be used to model two-dimensional plate structures.

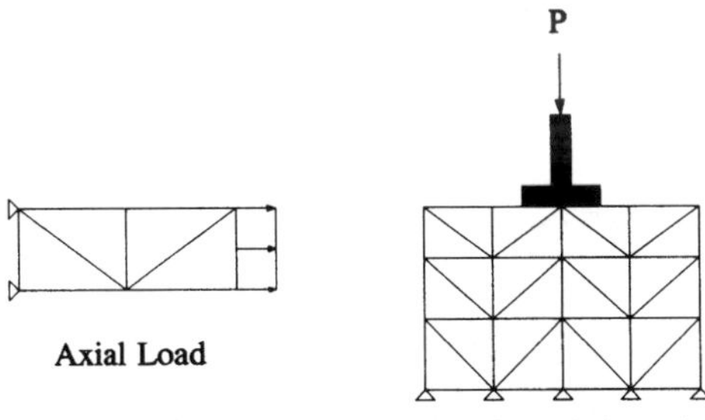

Figure 9-2

The basic procedures for combining elemental stiffnesses to create the global structural stiffness matrix, and then solving for nodal displacements and elemental stresses, are identical to those used for the elements discussed previously in this text. However, the derivation of the elemental stiffness matrix will make use of work and energy principles. Note that we will be using global coordinates, and as a result, coordinate transformation will be unnecessary.

9.2 PLANE STRESS AND PLANE STRAIN

There are two types of two-dimensional formulations for solving plane elasticity problems: plane stress and plane strain. To illustrate the assumptions associated with each formulation, consider Figure 9-3. We have assumed that the major dimensions of the thin elements lie in the x-y plane.

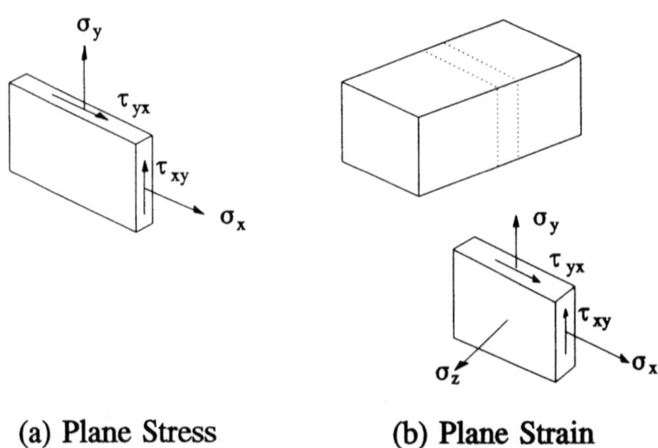

Figure 9-3 Plane stress and plane strain.

For the state of plane stress shown in Figure 9-3a, we consider a "free slice." That is, we assume that no stresses act in the z-direction, which is perpendicular to the plate. Thus, $\sigma_z = \tau_{xz} = \tau_{yz} = 0$.

The stress-strain relationships, or constitutive equations for plane stress assuming a linear, elastic, homogeneous, isotropic material, become (refer to your strength of materials text)

$$\sigma_x = \frac{E}{1-\nu^2}(\epsilon_x + \nu\epsilon_y) \tag{9.1}$$

$$\sigma_y = \frac{E}{1-\nu^2}(\epsilon_y + \nu\epsilon_x) \tag{9.2}$$

$$\tau_{xy} = G\gamma_{xy} \tag{9.3}$$

where

$$G = \frac{E}{2(1+\nu)} \tag{9.4}$$

Writing the above equations in matrix form, we have

$$\begin{Bmatrix} \sigma_x \\ \sigma_y \\ \tau_{xy} \end{Bmatrix} = \frac{E}{1-\nu^2}\begin{bmatrix} 1 & \nu & 0 \\ \nu & 1 & 0 \\ 0 & 0 & \frac{1-\nu}{2} \end{bmatrix}\begin{Bmatrix} \epsilon_x \\ \epsilon_y \\ \gamma_{xy} \end{Bmatrix} \tag{9.5}$$

or

$$\{\sigma\} = [C]_{stress}\{\epsilon\} \tag{9.6}$$

where $[C]_{stress}$ is the constitutive matrix for plane stress.

For the state of plane strain shown in Figure 9-3b, we consider a "constrained slice." That is, the strain in the z-direction is considered zero. Thus, $\sigma_z \neq 0$.

The stress-strain relationships for plane strain become

$$\sigma_x = \frac{E}{(1+\nu)(1-2\nu)}\left[(1-\nu)\epsilon_x + \nu\epsilon_y\right] \tag{9.7}$$

$$\sigma_y = \frac{E}{(1+\nu)(1-2\nu)}\left[\nu\epsilon_x + (1-\nu)\epsilon_y\right] \tag{9.8}$$

$$\tau_{xy} = G\gamma_{xy} \tag{9.9}$$

Also,

$$\sigma_z = \frac{\nu E}{(1+\nu)(1-2\nu)}(\epsilon_x + \epsilon_y) = \nu(\sigma_x + \sigma_y) \tag{9.10}$$

Writing the plane strain equations in matrix form yields

$$\begin{Bmatrix} \sigma_x \\ \sigma_y \\ \tau_{xy} \end{Bmatrix} = \frac{E}{(1+\nu)(1-2\nu)}\begin{bmatrix} 1-\nu & \nu & 0 \\ \nu & 1-\nu & 0 \\ 0 & 0 & \frac{1-2\nu}{2} \end{bmatrix}\begin{Bmatrix} \epsilon_x \\ \epsilon_y \\ \gamma_{xy} \end{Bmatrix} \tag{9.11}$$

or

$$\{\sigma\} = [C]_{strain}\{\epsilon\} \tag{9.12}$$

We will use the stress-strain relationships presented above later in this development to formulate the strain energy in terms of strain and, ultimately, displacements.

9.3 SHAPE FUNCTIONS FOR THE THREE-NODE TRIANGULAR ELEMENT

Since we will be deriving the structural stiffness equation by using the principle of minimum potential energy, we need to determine shape functions for the triangular element. Recall from Chapter 8 that the first step in this process is to assume a displacement function. Although several coordinate systems can be used, we shall use the global coordinate system shown in Figure 9-1. Note that we *must* assume a displacement function in this case since no closed-form force-displacement relationships can be derived directly from strength of materials and/or elementary structural theory.

Remember that the displacement field assumption we make must contain as many coefficients as degrees of freedom of the element. These coefficients will be expressed in terms of the nodal displacements, thus defining our shape functions. Since we have a three-node element with three degrees of freedom in each coordinate direction, we assume the following form for the displacement functions:

$$u(x, y) = a_1 + a_2 x + a_3 y \tag{9.13}$$

$$v(x, y) = b_1 + b_2 x + b_3 y \tag{9.14}$$

The above assumptions represent the simplest displacement functions possible for this element.

We next evaluate equation (9.13) at the nodal locations, giving

$$u_1 = a_1 + a_2 x_1 + a_3 y_1 \tag{9.15}$$

$$u_2 = a_1 + a_2 x_2 + a_3 y_2 \tag{9.16}$$

$$u_3 = a_1 + a_2 x_3 + a_3 y_3 \tag{9.17}$$

The above equations can be written

$$\begin{Bmatrix} u_1 \\ u_2 \\ u_3 \end{Bmatrix} = \begin{bmatrix} 1 & x_1 & y_1 \\ 1 & x_2 & y_2 \\ 1 & x_3 & y_3 \end{bmatrix} \begin{Bmatrix} a_1 \\ a_2 \\ a_3 \end{Bmatrix} \tag{9.18}$$

Similarly, equation (9.14) yields

$$\begin{Bmatrix} v_1 \\ v_2 \\ v_3 \end{Bmatrix} = \begin{bmatrix} 1 & x_1 & y_1 \\ 1 & x_2 & y_2 \\ 1 & x_3 & y_3 \end{bmatrix} \begin{Bmatrix} b_1 \\ b_2 \\ b_3 \end{Bmatrix} \tag{9.19}$$

We must now solve equations (9.18) and (9.19) for the coefficients a_1 through a_3 and b_1 through b_3 in terms of the nodal displacements.

After performing this task and back-substituting into equations (9.13) and (9.14) we find

$$u = [\,N_1 \quad N_2 \quad N_3\,] \begin{Bmatrix} u_1 \\ u_2 \\ u_3 \end{Bmatrix} \tag{9.20}$$

and

$$v = [\,N_1 \quad N_2 \quad N_3\,] \begin{Bmatrix} v_1 \\ v_2 \\ v_3 \end{Bmatrix} \tag{9.21}$$

where the shape functions N_1, N_2, and N_3 are given below.

$$N_1 = \frac{1}{D}[(x_2 y_3 - x_3 y_2) + (y_2 - y_3)x + (x_3 - x_2)y] \tag{9.22}$$

$$N_2 = \frac{1}{D}[(x_3y_1 - x_1y_3) + (y_3 - y_1)x + (x_1 - x_3)y] \tag{9.23}$$

$$N_3 = \frac{1}{D}[(x_1y_2 - x_2y_1) + (y_1 - y_2)x + (x_2 - x_1)y] \tag{9.24}$$

where

$$D = x_2y_3 - x_3y_2 + x_1(y_2 - y_3) + y_1(x_3 - x_2) \tag{9.25}$$

The expression for D is equal to twice the area of the triangular element. Combining equations (9.20) and (9.21) into a single matrix equation, we have

$$\begin{Bmatrix} u \\ v \end{Bmatrix} = \begin{bmatrix} N_1 & 0 & N_2 & 0 & N_3 & 0 \\ 0 & N_1 & 0 & N_2 & 0 & N_3 \end{bmatrix} \begin{Bmatrix} u_1 \\ v_1 \\ u_2 \\ v_2 \\ u_3 \\ v_3 \end{Bmatrix} \tag{9.26}$$

or

$$\{u\} = [N]\{q\} \tag{9.27}$$

9.4 STRAIN-DISPLACEMENT RELATIONSHIPS AND STRAIN ENERGY

From strength of materials, recall that the strain-displacement relationships for small strain can be written

$$\epsilon_x = \frac{\partial u}{\partial x}, \qquad \epsilon_y = \frac{\partial v}{\partial y}, \qquad \gamma_{xy} = \frac{\partial u}{\partial y} + \frac{\partial v}{\partial x} \tag{9.28}$$

or, in matrix form,

$$\begin{Bmatrix} \epsilon_x \\ \epsilon_y \\ \gamma_{xy} \end{Bmatrix} = \begin{bmatrix} \frac{\partial}{\partial x} & 0 \\ 0 & \frac{\partial}{\partial y} \\ \frac{\partial}{\partial y} & \frac{\partial}{\partial x} \end{bmatrix} \begin{Bmatrix} u \\ v \end{Bmatrix} \tag{9.29}$$

Using equation (9.26) in equation (9.29) and performing the indicated differentiations, we obtain

$$\begin{Bmatrix} \epsilon_x \\ \epsilon_y \\ \gamma_{xy} \end{Bmatrix} = \frac{1}{D} \begin{bmatrix} (y_2 - y_3) & 0 & (y_3 - y_1) & 0 & (y_1 - y_2) & 0 \\ 0 & (x_3 - x_2) & 0 & (x_1 - x_3) & 0 & (x_2 - x_1) \\ (x_3 - x_2) & (y_2 - y_3) & (x_1 - x_3) & (y_3 - y_1) & (x_2 - x_1) & (y_1 - y_2) \end{bmatrix} \begin{Bmatrix} u_1 \\ v_1 \\ u_2 \\ v_2 \\ u_3 \\ v_3 \end{Bmatrix} \tag{9.30}$$

or

$$\{\epsilon\} = [B]\{q\} \tag{9.31}$$

where D is again given by equation (9.25).

Note that the three-node triangular element is a constant strain element. This is a direct result of the assumed displacement fields and the strain-displacement relationships.

We next formulate the expression for strain energy.

In Chapter 8 we expressed the strain energy in equation (8.73), which is repeated here for convenience.

$$U = \int_V \frac{1}{2} \{\epsilon\}^T \{\sigma\} \, dV \tag{8.73}$$

Using equation (9.6) or (9.12) in the above equation yields

$$U = \int_V \frac{1}{2} \{\epsilon\}^T [C] \{\epsilon\} \, dV \tag{9.32}$$

From equation (9.31),

$$\{\epsilon\}^T = \{q\}^T [B]^T \tag{9.33}$$

Thus, equation (9.32) becomes

$$U = \int_V \frac{1}{2} \{q\}^T [B]^T [C] [B] \{q\} \, dV \tag{9.34}$$

Assuming the element has constant thickness t, then $dV = t \, dA$. Also note that $\{q\}$, $[B]$, and $[C]$ contain elements that are constants.

Thus,

$$\begin{aligned} U &= \frac{1}{2} \{q\}^T [B]^T [C] [B] \{q\} t \int_A dA \\ &= \frac{1}{2} \{q\}^T [B]^T [C] [B] \{q\} t A \end{aligned} \tag{9.35}$$

where A is the area of the element and equals $D/2$ (equation [9.25]).

9.5 FORCE POTENTIAL

We shall consider only applied nodal forces in this brief introduction to the finite element method. Distributed forces (tractions) on the edges of the element and body forces would need to be included for completeness. These are left for a later course in the finite element method.

The force potential of nodal forces was presented in Chapter 8. For a force at node 1 with components F_{1x} and F_{1y}, the force potential is

$$V_1 = -F_{1x} u_1 - F_{1y} v_1 \tag{9.36}$$

For forces at all three nodes,

$$V_{nodal\ forces} = -\{q\}^T \begin{Bmatrix} F_{1x} \\ F_{1y} \\ F_{2x} \\ F_{2y} \\ F_{3x} \\ F_{3y} \end{Bmatrix}$$

$$= -[u_1 \quad v_1 \quad u_2 \quad v_2 \quad u_3 \quad v_3] \begin{Bmatrix} F_{1x} \\ F_{1y} \\ F_{2x} \\ F_{2y} \\ F_{3x} \\ F_{3y} \end{Bmatrix} \tag{9.37}$$

9.6 APPLICATION OF THE PRINCIPLE OF MINIMUM POTENTIAL ENERGY

Summing the strain energy from equation (9.34) and the nodal force potential from equation (9.37), we find the total potential energy to be

$$\Pi = U + V_{nodal\ forces}$$

$$= \int_V \frac{1}{2} \{q\}^T [B]^T [C][B]\{q\}\, dV - \{q\}^T \begin{Bmatrix} F_{1x} \\ F_{1y} \\ F_{2x} \\ F_{2y} \\ F_{3x} \\ F_{3y} \end{Bmatrix} \tag{9.38}$$

We next differentiate the potential energy with respect to each displacement, which is equivalent to differentiating with respect to the nodal displacements $\{q\}$. Thus, we find

$$\int_V [B]^T [C][B]\, dV \begin{Bmatrix} u_1 \\ v_1 \\ u_2 \\ v_2 \\ u_3 \\ v_3 \end{Bmatrix} = \begin{Bmatrix} F_{1x} \\ F_{1y} \\ F_{2x} \\ F_{2y} \\ F_{3x} \\ F_{3y} \end{Bmatrix} \tag{9.39}$$

or

$$[k]\{q\} = \{F\}_{nodal\ forces} \tag{9.40}$$

For this element,

$$[k] = [B]^T [C][B]\, tA \tag{9.41}$$

since $[B]$, $[C]$, and t are constant.

We next consider a structure modeled with three-node triangular elements as an example problem.

9.7 EXAMPLE PLANE STRESS PROBLEM

A shearwall is modeled as shown in Figure 9-4.

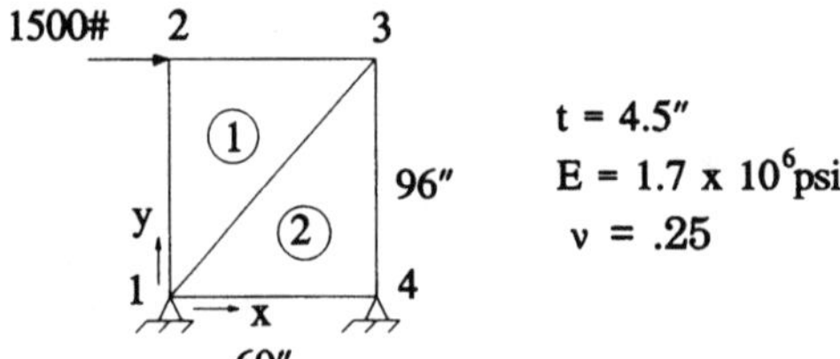

Figure 9-4 Example problem.

Assuming plane stress, solve for the nodal displacements and stresses. The nodes of each element are numbered in sequence counterclockwise.

Element 1:

$$x_1 = y_1 = 0$$

$$x_2 = 60'', \qquad y_2 = 96''$$

$$x_3 = 0, \qquad y_3 = 96''$$

From equation (9.25),

$$D = 60(96) - 0 + 0 + 0 = 5760 \text{ in}^2 = 2 \times \text{Area}$$

From equation (9.30),

$$[B] = \frac{1}{5760}\begin{bmatrix} 0 & 0 & 96 & 0 & -96 & 0 \\ 0 & -60 & 0 & 0 & 0 & 60 \\ -60 & 0 & 0 & 96 & 60 & -96 \end{bmatrix} \tag{9.42}$$

From equation (9.5),

$$[C]_{stress} = \frac{1.7 \times 10^6}{.9375}\begin{bmatrix} 1 & .25 & 0 \\ .25 & 1 & 0 \\ 0 & 0 & .375 \end{bmatrix} \tag{9.43}$$

Equation (9.41) yields

$$[k]_1 = 708.333 \begin{array}{c} \begin{array}{cccccc} \text{①} & \text{②} & \text{⑤} & \text{⑥} & \text{③} & \text{④} \end{array} \\ \begin{bmatrix} 1350 & 0 & 0 & -2160 & -1350 & 2160 \\ 0 & 3600 & -1440 & 0 & 1440 & -3600 \\ 0 & -1440 & 9216 & 0 & -9216 & 1440 \\ -2160 & 0 & 0 & 3456 & 2160 & -3456 \\ -1350 & 1440 & -9216 & 2160 & 10566 & -3600 \\ 2160 & -3600 & 1440 & -3456 & -3600 & 7056 \end{bmatrix} \end{array} \begin{array}{c} \text{①} \\ \text{②} \\ \text{⑤} \\ \text{⑥} \\ \text{③} \\ \text{④} \end{array} \tag{9.44}$$

Element 2:

$$x_1 = y_1 = 0$$

$$x_2 = 60'', \qquad y_2 = 0$$

$$x_3 = 60'', \qquad y_3 = 96''$$

From equation (9.25)

$$D = 60(96) - 0 + 0 + 0 = 5760 \text{ in}^2$$

From equation (9.30),

$$[B] = \frac{1}{5760}\begin{bmatrix} -96 & 0 & 96 & 0 & 0 & 0 \\ 0 & 0 & 0 & -60 & 0 & 60 \\ 0 & -96 & -60 & 96 & 60 & 0 \end{bmatrix} \tag{9.45}$$

From equation (9.5),

$$[C]_{stress} = \frac{1.7 \times 10^6}{.9375}\begin{bmatrix} 1 & .25 & 0 \\ .25 & 1 & 0 \\ 0 & 0 & .375 \end{bmatrix} \tag{9.46}$$

Equation (9.41) yields

$$[k]_2 = 708.333 \begin{bmatrix} 9216 & 0 & -9216 & 1440 & 0 & -1440 \\ 0 & 3456 & 2160 & -3456 & -2160 & 0 \\ -9216 & 2160 & 10566 & -3600 & -1350 & 1440 \\ 1440 & -3456 & -3600 & 7056 & 2160 & -3600 \\ 0 & -2160 & -1350 & 2160 & 1350 & 0 \\ -1440 & 0 & 1440 & -3600 & 0 & 3600 \end{bmatrix} \tag{9.47}$$

(columns ① ② ⑦ ⑧ ⑤ ⑥; rows ① ② ⑦ ⑧ ⑤ ⑥)

Combining, accounting for zero boundary conditions,

$$[K] = 708.333 \begin{bmatrix} 10566 & -3600 & -9216 & 2160 \\ -3600 & 7056 & 1440 & -3456 \\ -9216 & 1440 & 10566 & 0 \\ 2160 & -3456 & 0 & 7056 \end{bmatrix} \tag{9.48}$$

(columns ③ ④ ⑤ ⑥; rows ③ ④ ⑤ ⑥)

Our overall structural stiffness equation becomes

$$[K]\begin{Bmatrix} u_3 \\ u_4 \\ u_5 \\ u_6 \end{Bmatrix} = \begin{Bmatrix} 1500\# \\ 0 \\ 0 \\ 0 \end{Bmatrix} \tag{9.49}$$

Inverting $[K]$ and solving for displacements, we find

$$\begin{Bmatrix} u_3 \\ u_4 \\ u_5 \\ u_6 \end{Bmatrix} = \begin{Bmatrix} .001370 \\ .000341 \\ .001148 \\ -.000252 \end{Bmatrix} in \tag{9.50}$$

The strains are given by equation (9.30).

Element 1:

$$\begin{Bmatrix} \epsilon_x \\ \epsilon_y \\ \gamma_{xy} \end{Bmatrix} = \frac{1}{5760}\begin{bmatrix} 0 & 0 & 96 & 0 & -96 & 0 \\ 0 & -60 & 0 & 0 & 0 & 60 \\ -60 & 0 & 0 & 96 & 60 & -96 \end{bmatrix}\begin{Bmatrix} 0 \\ 0 \\ .001148 \\ -.000252 \\ .001370 \\ .000341 \end{Bmatrix}$$

$$= \begin{Bmatrix} -3.70 \times 10^{-6}\ in/in \\ 3.55 \times 10^{-6}\ in/in \\ 33.41 \times 10^{-6}\ rad \end{Bmatrix} \tag{9.51}$$

Element 2:

$$\begin{Bmatrix} \epsilon_x \\ \epsilon_y \\ \gamma_{xy} \end{Bmatrix} = \frac{1}{5760}\begin{bmatrix} -96 & 0 & 96 & 0 & 0 & 0 \\ 0 & 0 & 0 & -60 & 0 & 60 \\ 0 & -96 & -60 & 96 & 60 & 0 \end{bmatrix}\begin{Bmatrix} 0 \\ 0 \\ 0 \\ 0 \\ .001148 \\ -.000252 \end{Bmatrix}$$

$$= \begin{Bmatrix} 0\ in/in \\ -2.63 \times 10^{-6}\ in/in \\ 11.96 \times 10^{-6}\ rad \end{Bmatrix} \tag{9.52}$$

The stresses are next calculated using equation (9.5).

Element 1:

$$\begin{Bmatrix} \sigma_x \\ \sigma_y \\ \tau_{xy} \end{Bmatrix} = \frac{1.7 \times 10^6}{.9375}\begin{bmatrix} 1 & .25 & 0 \\ .25 & 1 & 0 \\ 0 & 0 & .375 \end{bmatrix}\begin{Bmatrix} -3.70 \times 10^{-6}\ in/in \\ 3.55 \times 10^{-6}\ in/in \\ 33.41 \times 10^{-6}\ rad \end{Bmatrix}$$

$$= \begin{Bmatrix} -5.10 \\ 12.87 \\ 22.72 \end{Bmatrix}\ psi \tag{9.53}$$

Element 2:

$$\begin{Bmatrix} \sigma_x \\ \sigma_y \\ \tau_{xy} \end{Bmatrix} = \frac{1.7 \times 10^6}{.9375}\begin{bmatrix} 1 & .25 & 0 \\ .25 & 1 & 0 \\ 0 & 0 & .375 \end{bmatrix}\begin{Bmatrix} 0\ in/in \\ -2.63 \times 10^{-6}\ in/in \\ 11.96 \times 10^{-6}\ rad \end{Bmatrix}$$

$$= \begin{Bmatrix} -1.19 \\ -4.77 \\ 8.13 \end{Bmatrix}\ psi \tag{9.54}$$

Keep in mind that modeling the shearwall with only two elements will not result in an accurate solution. Many more elements would have to be used to obtain a realistic solution. Naturally, the problem will then become algebraically intractable for hand solution, and a computer solution will be required.

9.8 SUMMARY

In this chapter we briefly presented an introduction to the finite element method. In doing so, we used the simplest element that requires work and energy principles for formulation of the elemental stiffness equation. Because of the simplicity of the three-node triangular element we were able to integrate the required stiffness equation in closed form. With more complicated elements such as the four-node quadrilateral and the six-node triangle, numerical integration is generally used to determine the elemental stiffnesses.

It is hoped that this chapter has kindled your interest in pursuing further study of the finite element method.

PROBLEMS

9.1 Solve explicitly for the a_1 through a_3 coefficients in equation (9.18).

9.2 Verify the expressions for the shape functions presented in equations (9.22) through (9.24).

9.3 Find the displacements and stresses for the plate shown in Figure P9-3. Assume a state of plane stress.

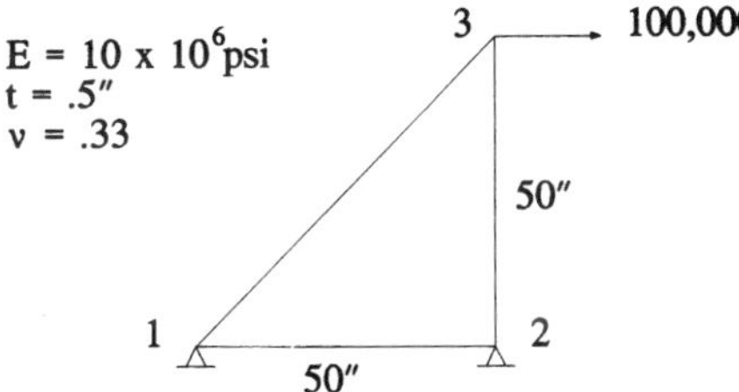

Figure P9-3

9.4 The plate in problem 9.3 is modeled with two elements as shown in Figure P9-4. Find the displacements and stresses.

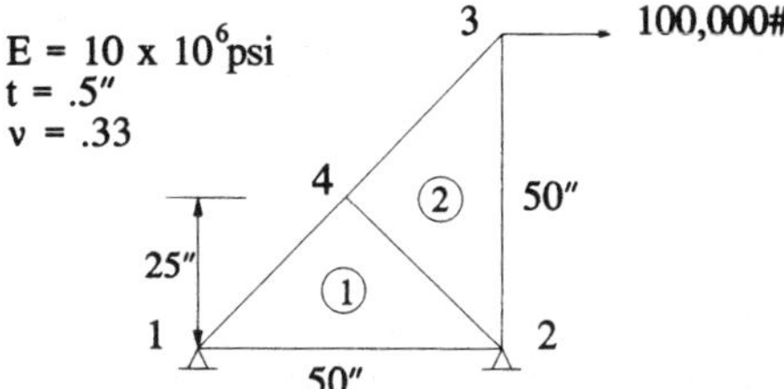

Figure P9-4

9.5 An axially loaded member is modeled as shown in Figure P9-5. Find displacements and stresses. Compare your results with the strength of materials solution for a concentrically loaded axial force member with the same total axial force.

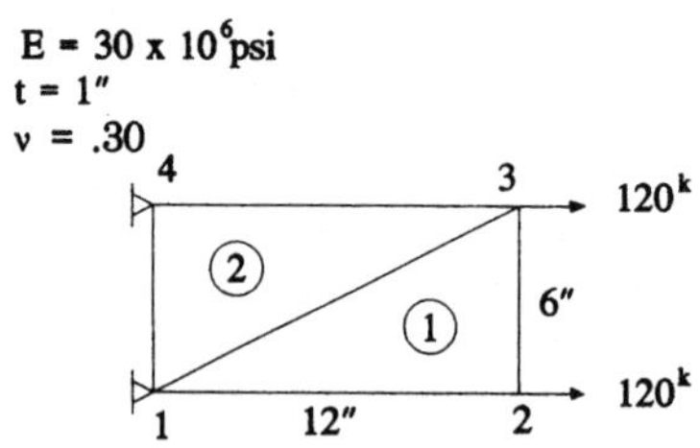

Figure P9-5

9.6 Solve for stresses and displacements of the plane strain structure shown in Figure P9-6.

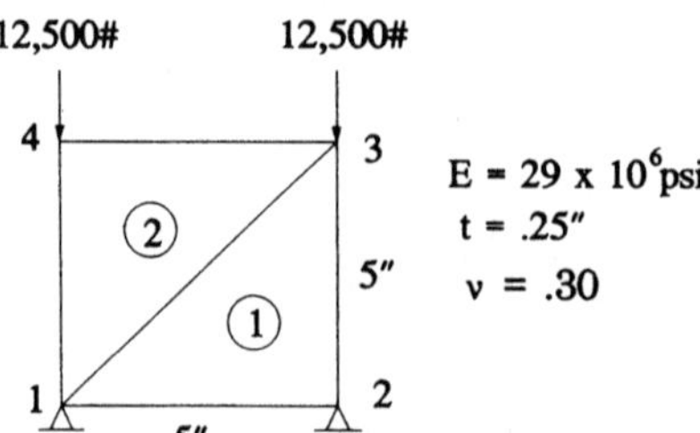

Figure P9-6

APPENDIX A

REVIEW OF MATRIX ALGEBRA

A.1 BASIC OPERATIONS

The four basic operations of matrix algebra are

(1) Addition
(2) Subtraction
(3) Multiplication
(4) Inversion (analogous to division)

A.2 BASIC DEFINITIONS

If m equals the number of rows and n equals the number of columns of a matrix, and if a_{ij} is the element of the matrix in row i and column j, we define the following types of matrices:

(a) Square Matrix: $m = n$
(b) Symmetric Matrix: $a_{ij} = a_{ji}$
(c) Diagonal Matrix: A square matrix with $a_{ij} = 0$ for $i \neq j$, $a_{ij} \neq 0$ for $i = j$
(d) Unit Matrix: A diagonal matrix with $a_{ii} = 1$, $i = 1, 2, 3 \ldots n$
(e) Column Matrix: An $m \times 1$ matrix designated by {}
(f) Row Matrix: A $1 \times n$ matrix designated by [ ]

Transpose of a matrix The transpose of a matrix $[A]$ is a matrix $[A]^T$ that is obtained by interchanging the rows and columns of the original matrix $[A]$. The matrix $[A]$ can be of any order.

For example:

$$\text{If } [A] = \begin{bmatrix} 1 & 3 & 2 \\ -4 & 0 & 5 \end{bmatrix} \quad \text{then} \quad [A]^T = \begin{bmatrix} 1 & -4 \\ 3 & 0 \\ 2 & 5 \end{bmatrix}$$

Thus, $a_{ij}^T = a_{ji}$

A.3 EXAMPLES OF OPERATIONS

Addition:
The matrices $[a]$ and $[b]$ must be of the same order $m \times n$ (same number of rows and same number of columns).

Then, the elements of the resulting matrix $[c] = [a] + [b]$ are as follows:

$$c_{ij} = a_{ij} + b_{ij}$$

Subtraction:
As in addition,

$$[c] = [a] - [b] \text{ and } c_{ij} = a_{ij} - b_{ij}$$

Scalar Multiplication:
If $[c] = K[a]$ where K is a scalar constant, then $c_{ij} = Ka_{ij}$

Matrix Multiplication:
The matrices being multiplied must be conformable. That is, the number of columns of the first matrix must be equal to the number of rows of the second matrix.

If $[a]$ is of order $m \times r$, and $[b]$ of order $r \times n$, the resulting product $[c]$ will be of order $m \times n$ where the elements are given by

$$c_{ij} = \sum_{k=1}^{r} a_{ik} b_{kj}$$

Example:

$$\begin{bmatrix} 2 & 3 & 6 \\ 4 & 1 & 0 \end{bmatrix} \begin{bmatrix} 1 & 1 & 3 \\ 2 & 5 & 9 \\ 0 & 1 & 0 \end{bmatrix} = \begin{bmatrix} 8 & 23 & 33 \\ 6 & 9 & 21 \end{bmatrix}$$

Note that $[a][b] \neq [b][a]$ (not commutative) even in the case of two square matrices. Therefore we must differentiate between pre- and postmultiplication.

Inversion:
We define the inverse $[a]^{-1}$ of a matrix $[a]$ to have the following property:

$$[a]^{-1}[a] = [a][a]^{-1} = [U] \text{ where } [U] \text{ is a unit matrix.}$$

Thus, the matrix to be inverted must be a square matrix. The rows and columns must also be independent—that is, the determinant of the matrix $[a]$ must not be zero.

The *cofactor* of an element of a matrix is defined as the signed minor of the element—that is, $(-1)^{i+j}$ times the minor. The minor of an element is obtained by evaluating the determinant that results from removing the row and column corresponding to that element.

If $[A]$ is the matrix of cofactors of $[a]$, then the *adjoint* of $[a] = [A]^T$.
The inverse of $[a]$ is given by $[a]^{-1} = adjoint$ of $[a]/|a|$.

Example:

$$[a] = \begin{bmatrix} 1 & -2 & 3 \\ 2 & 0 & -3 \\ 1 & 1 & 1 \end{bmatrix}$$

$|a| = 1(3) - 2(-5) + 1(6) = 19 \neq 0$, thus the inverse exists.

$$cofactor\ [a] = [A] = \begin{bmatrix} 3 & -5 & 2 \\ 5 & -2 & 3 \\ 6 & 9 & 4 \end{bmatrix}$$

$$adjoint\ [a] = [A]^T = \begin{bmatrix} 3 & 5 & 6 \\ -5 & -2 & 9 \\ 2 & -3 & 4 \end{bmatrix}$$

Thus,

$$[a]^{-1} = \frac{1}{19}\begin{bmatrix} 3 & 5 & 6 \\ -5 & -2 & 9 \\ 2 & -3 & 4 \end{bmatrix}$$

Check:

$$[a][a]^{-1} = \frac{1}{19}\begin{bmatrix} 1 & -2 & 3 \\ 2 & 0 & -3 \\ 1 & 1 & 1 \end{bmatrix}\begin{bmatrix} 3 & 5 & 6 \\ -5 & -2 & 9 \\ 2 & -3 & 4 \end{bmatrix} = \frac{1}{19}\begin{bmatrix} 19 & 0 & 0 \\ 0 & 19 & 0 \\ 0 & 0 & 19 \end{bmatrix} = \begin{bmatrix} 1 & 0 & 0 \\ 0 & 1 & 0 \\ 0 & 0 & 1 \end{bmatrix}$$

Many methods can be used to invert matrices, of which the previous procedure is just one. For most computer work, a modified Gauss-Jordan method is often used.

EXERCISES

#1) Evaluate

$$[X] = \{[A]^{-1}[B] - [E]^{-1}[D]\}^{-1}[A]^{-1}$$

$$[Y] = \{[D]^{-1}[E] - [B]^{-1}[A]\}^{-1}[D]^{-1}$$

where

$$[A] = \begin{bmatrix} 0 & 7 \\ 1 & 5 \end{bmatrix} [B] = \begin{bmatrix} 1 & 2 \\ 2 & 2 \end{bmatrix} [D] = \begin{bmatrix} 9 & 4 \\ 6 & 1 \end{bmatrix} [E] = \begin{bmatrix} 8 & 3 \\ 0 & 1 \end{bmatrix}$$

#2) Given

$$[A]\{x\} = [B]\{y\} + \{C\} \text{ and } [D]\{y\} = [E]\{z\} + \{F\};$$

find and evaluate $\{x\} = [G]\{z\} + \{H\}$ where

$$[A] = \begin{bmatrix} 6 & 3 \\ 4 & 1 \end{bmatrix} [B] = \begin{bmatrix} 3 & 1 \\ 1 & 4 \end{bmatrix} \{C\} = \begin{Bmatrix} 1 \\ 4 \end{Bmatrix}$$

$$[D] = \begin{bmatrix} 4 & 3 \\ 5 & 1 \end{bmatrix} [E] = \begin{bmatrix} 7 & 8 \\ 6 & 5 \end{bmatrix} \{F\} = \begin{Bmatrix} 8 \\ 9 \end{Bmatrix}$$

#3) Given

$[Q] = [B_1]\{[I] + [B_2 - B_1]^{-1}[B_1]\}[P]$ and

$[Q_1] = [B_2]\{[B_2 - B_1]^{-1}[B_1]\}[P]$, prove that

$[Q] = [Q_1]$

APPENDIX B

MATRIX INVERSION ROUTINES

Appendix B presents two matrix inversion routines. The first will invert any matrix, symmetric or not. The second requires the matrix to be symmetric and stored in half-bandwidth form. The force and restraint code matrices are also required in the second listing since this routine calculates displacements directly.

Routine 1:

```
'Matrix inversion routine
'skr is the matrix to be inverted
'and is destroyed in the process.
'After inversion skr contains the inverse,
'so make sure you save skr in a different array before
'calling this routine if you need it later.
'nkr is the order of the matrix.
FOR K=1 TO NKR
AM=0
100 FOR I=K TO NKR
FOR J=K TO NKR
IF(ABS(AM))>ABS(SKR(I,J)) THEN GOTO 160
AM=SKR(I,J)
IK(K)=I
JK(K)=J
160 NEXT J:NEXT I
IF AM<>0 THEN GOTO 190
PRINT ''zero MATRIX***ABORT":END
190 I=IK(K)
IF I=K THEN GOTO 260
IF I<K THEN GOTO 100
FOR J=1 TO NKR
ST=SKR(K,J)
SKR(K,J)=SKR(I,J)
SKR(I,J)=-ST:NEXT J
260 J=JK(K)
IF J<K THEN GOTO 100
IF J=K THEN GOTO 330
FOR I=1 TO NKR
ST=SKR(I,K)
SKR(I,K)=SKR(I,J)
SKR(I,J)=-ST:NEXT I
330 FOR I=1 TO NKR
```

```
IF I=K THEN GOTO 360
SKR(I,K)=-SKR(I,K)/AM
360 NEXT I
FOR I=1 TO NKR
FOR J=1 TO NKR
IF I=K THEN GOTO 420
IF J=K THEN GOTO 420
SKR(I,J)=SKR(I,J)+SKR(I,K)*SKR(K,J)
420 NEXT J:NEXT I
FOR J=1 TO NKR
IF J=K THEN GOTO 460
SKR(K,J)=SKR(K,J)/AM
460 NEXT J
SKR(K,K)=1/AM
NEXT K
FOR L=1 TO NKR
K=NKR-L+1
J=IK(K)
IF J<=K THEN GOTO 570
FOR I=1 TO NKR
ST=SKR(I,K)
SKR(I,K)=-SKR(I,J)
SKR(I,J)=ST:NEXT I
570 I= JK(K)
IF I<=K THEN GOTO 630
FOR J=1 TO NKR
ST=SKR(K,J)
SKR(K,J)=-SKR(I,J)
SKR(I,J)=ST:NEXT J
630 NEXT L
RETURN
```

Routine 2:

Assume that the stiffness matrix is stored in half-bandwidth form as shown in section 7.1.

Let DOF = number of degrees of freedom per node.

We first construct a one-dimensional array of boundary constraints. Call this array BCOND(I)

```
KSUM = 0
FOR I = 1 TO NN  'loop on number of nodes
KSUM = KSUM + 1
BCOND(KSUM) = KXRES(I)
KSUM = KSUM + 1
BCOND(KSUM) = KYRES(I)
```

```
...     'Continue until all restraints have been accounted for.
        'For example, the 2-D frame will have KXRES(I),KYRES(I),
        ' and KZRES(I) for each node.
NEXT I
```

Assume that DU(I) (the global displacement matrix) contains all zeros except for specified non-zero support movements.

We duplicate the nodal force matrix F(I) in an array FD(I) in order to save the actual applied loads since F(I) will be modified during the solution for displacements.

For non-zero boundary conditions we modify the main diagonal term of the stiffness matrix and the force matrix as described in section 7.1.

```
N = DOF*NN
FOR I=1 TO N
FD(I) = F(I)
IF BCOND(I) = 0 THEN 10
SK(I,1)=SK(I,1)*1E9
F(I)=DU(I)*SK(I,1)
10 NEXT I
```

We now begin the solution routine.

```
FOR I=1 TO N
IF SK(I,1)=0 THEN 40
FOR J=2 TO BW 'BW = half-bandwidth (calculated previously)
IF SK(I,J)=0 THEN 30
L=I+J-1
RATIO = SK(I,J)/SK(I,1)
JJ=0
FOR K=J TO BW
JJ=JJ+1
IF L>N THEN 20
SK(L,JJ)=SK(L,JJ)-RATIO*SK(I,K)
20 NEXT K
SK(I,J)=RATIO
30 NEXT J
40 NEXT I
FOR I=1 TO N
IF SK(I,1)=0 THEN 60
FOR J=2 TO BW
IF SK(I,J)=0 THEN 50
L=I+J-1
IF L>N THEN 50
F(L)=F(L)-SK(I,J)*F(I)
50 NEXT J
F(I)=F(I)/SK(I,1)
60 NEXT I
```

```
DU(N)=F(N)
FOR LL=2 TO N
I=N-LL+1
FOR J=2 TO BW
IF SK(I,J)=0 THEN 70
K=I+J-1
IF K>N THEN 80
F(I)=F(I)-SK(I,J)*DU(K)
70 NEXT J
DU(I)=F(I)  'DU contains the final displacements
80 NEXT LL
```

We can now restore the original forces into the F(I) matrix.

```
FOR I=1 TO N
F(I)=FD(I)
NEXT I
```

APPENDIX C

DERIVATION OF THE SLOPE-DEFLECTION EQUATIONS

Consider a simply supported beam ab subjected to end moments M_{ab} and M_{ba}, applied loads $p(x)$, and a support movement Δ. All forces and displacements, including end rotations θ_a and θ_b, are shown in their positive directions in Figure APPC-1.

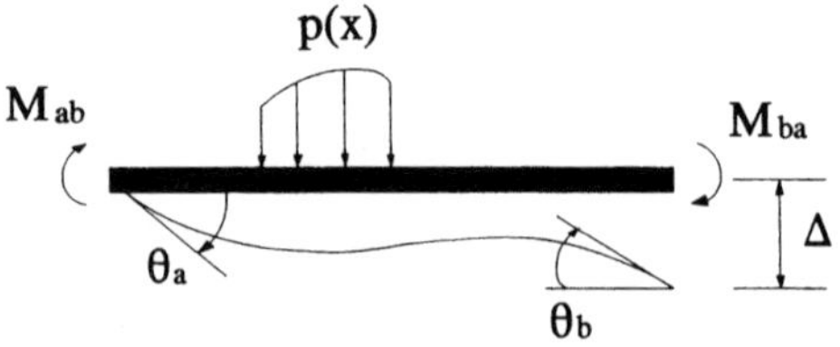

Figure APPC-1 Positive forces and displacements.

Note that positive moments and rotations are clockwise, and Δ is positive when it causes a clockwise rotation of the chord of the beam.

We shall write expressions for the total rotations at the ends of the member owing to the effects of all loads and support movements.

$$\theta_a = \theta_a^{M_{ab}} + \theta_a^{M_{ba}} + \theta_a^{loads} + \theta_a^{\Delta} \tag{APPC.1}$$

$$\theta_b = \theta_b^{M_{ab}} + \theta_b^{M_{ba}} + \theta_b^{loads} + \theta_b^{\Delta} \tag{APPC.2}$$

Assume that EI = constant. Many techniques can be used to find the displacements in equations (APPC.1) and (APPC.2). These include moment area, virtual work, double integration, and conjugate beam.

Solving for the rotation angles θ_a and θ_b due to the applied moments M_{ab} and M_{ba} we find,

$$\theta_a^{Mab} = M_{ab}l/3EI \qquad \theta_b^{Mab} = -M_{ab}l/6EI$$

$$\theta_a^{Mba} = -M_{ba}l/6EI \qquad \theta_b^{Mba} = M_{ba}l/3EI$$

For the support movement Δ, if we assume small displacements, tan $\theta \approx \theta$ and we have

$$\theta_a^{\Delta} = \Delta/l \qquad \theta_b^{\Delta} = \Delta/l$$

Equations (APPC.1) and (APPC.2) become

$$\theta_a = \frac{M_{ab}l}{3EI} - \frac{M_{ba}l}{6EI} + \theta_a^{loads} + \frac{\Delta}{l} \tag{APPC.3}$$

$$\theta_b = -\frac{M_{ab}l}{6EI} + \frac{M_{ba}l}{3EI} + \theta_b^{loads} + \frac{\Delta}{l} \tag{APPC.4}$$

Solving equations (APPC.3) and (APPC.4) for M_{ab} and M_{ba} we find

$$M_{ab} = \frac{2EI}{l}\left[2\theta_a + \theta_b - \frac{3\Delta}{l}\right] - \frac{2EI}{l}\left[2\theta_a^{loads} + \theta_b^{loads}\right] \quad \text{(APPC.5)}$$

$$M_{ba} = \frac{2EI}{l}\left[\theta_a + 2\theta_b - \frac{3\Delta}{l}\right] - \frac{2EI}{l}\left[2\theta_b^{loads} + \theta_a^{loads}\right] \quad \text{(APPC.6)}$$

Equations (APPC.5) and (APPC.6) are the general form of the slope-deflection equations.

Consider the case when $\theta_a = \theta_b = \Delta = 0$. Then,

$$M_{ab} = -\frac{2EI}{l}\left[2\theta_a^{loads} + \theta_b^{loads}\right] \quad \text{(APPC.7)}$$

$$M_{ba} = -\frac{2EI}{l}\left[2\theta_b^{loads} + \theta_a^{loads}\right] \quad \text{(APPC.8)}$$

Now the support conditions we have specified are those of a fixed-ended beam. Thus, the moments in equations (APPC.7) and (APPC.8) are called the *fixed end moments.* Note that these moments can be found by computing the rotations at the ends of a *simply supported beam* due to the loads applied to the beam.

The slope-deflection equations can now be written

$$M_{ab} = \frac{2EI}{l}\left[2\theta_a + \theta_b - \frac{3\Delta}{l}\right] \pm M_{ab}^{fixed} \quad \text{(APPC.9)}$$

$$M_{ba} = \frac{2EI}{l}\left[2\theta_b + \theta_a - \frac{3\Delta}{l}\right] \pm M_{ba}^{fixed} \quad \text{(APPC.10)}$$

Keep in mind that the fixed end moments will be positive when acting clockwise on the beam.

We can write these equations in even a more compact form. If the loads are zero and θ_a and θ_b are also zero, we have

$$M_{ab} = -6EI\Delta/l^2 \qquad M_{ba} = -6EI\Delta/l^2$$

The above expressions are the fixed end moments due to support movements. Thus,

$$M_{ab} = \frac{2EI}{l}[2\theta_a + \theta_b] \pm M_{ab}^{fixed} \quad \text{(APPC.11)}$$

$$M_{ba} = \frac{2EI}{l}[2\theta_b + \theta_a] \pm M_{ba}^{fixed} \quad \text{(APPC.12)}$$

where the fixed end moments in equations (APPC.11) and (APPC.12) include the effects of loads and support movements, positive clockwise at the ends of the member.

The slope-deflection equations used in concert with joint equilibrium equations can be used to solve many structural problems involving beams and frames. The number of simultaneous equations that must be solved will be equal to the number of kinematic degrees of freedom. Thus, the method is a displacement method of analysis.

Example

Consider the beam shown in Figure APPC-2.

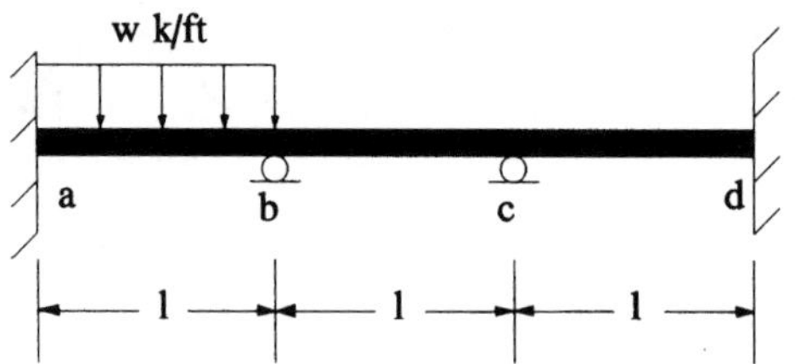

Figure APPC-2

The fixed end moments for a beam with a uniformly distributed load applied are given by $\pm wl^2/12$. Thus,

$$M_{ab}^{fixed} = -wl^2/12 \text{ and } M_{ba}^{fixed} = wl^2/12$$

Writing the slope-deflection equations for each member we have

Member *ab*:

$$M_{ab} = (2EI/l)[2\theta_a + \theta_b] - wl^2/12 = (2EI/l)\theta_b - wl^2/12$$

$$M_{ba} = (2EI/l)[2\theta_b + \theta_a] + wl^2/12 = (4EI/l)\theta_b + wl^2/12$$

Member *bc*:

$$M_{bc} = (2EI/l)[2\theta_b + \theta_c]$$

$$M_{cb} = (2EI/l)[2\theta_c + \theta_b]$$

Member *cd*:

$$M_{cd} = (2EI/l)[2\theta_c + \theta_d] = (4EI/l)\theta_c$$

$$M_{dc} = (2EI/l)[2\theta_d + \theta_c] = (2EI/l)\theta_c$$

Figure APPC-3 shows free-body diagrams of joints b and c.

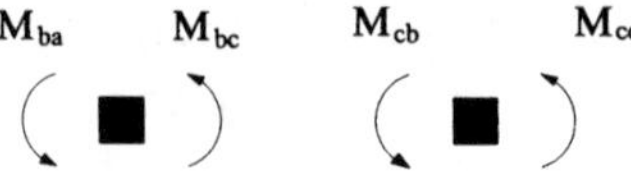

Figure APPC-3 Free-body diagrams of joints.

Thus,

$$M_{ba} + M_{bc} = 0$$

$$M_{cb} + M_{cd} = 0$$

Substituting the slope-deflection equations into the above equilibrium equations yields

$$(4EI/l)\theta_b + wl^2/12 + (2EI/l)[2\theta_b + \theta_c] = 0$$

or

$$\frac{8EI}{l}\theta_b + \frac{2EI}{l}\theta_c = -\frac{wl^2}{12} \tag{APPC.13}$$

and

$$(2EI/l)[2\theta_c + \theta_b] + (4EI/l)\theta_c = 0$$

or

$$\frac{2EI}{l}\theta_b + \frac{8EI}{l}\theta_c = 0 \tag{APPC.14}$$

Solving equations (APPC.13) and (APPC.14) for the rotations θ_b and θ_c we find

$$\theta_b = -\frac{wl^3}{90EI} \qquad \theta_c = \frac{wl^3}{360EI} \tag{APPC.15}$$

Back-substituting the rotation angles into the slope-deflection equations for each member yields the final moments at the ends of the members.

$$M_{ab} = -.1055\ wl^2$$

$$M_{ba} = .0389\ wl^2 = -M_{bc}$$

$$M_{cb} = -.0111\ wl^2 = -M_{cd}$$

$$M_{dc} = .0055\ wl^2$$

The end shears and reactions can now be found by drawing free-body diagrams of the members and joints and writing equilibrium equations. These free-body diagrams are shown in Figure APPC-4.

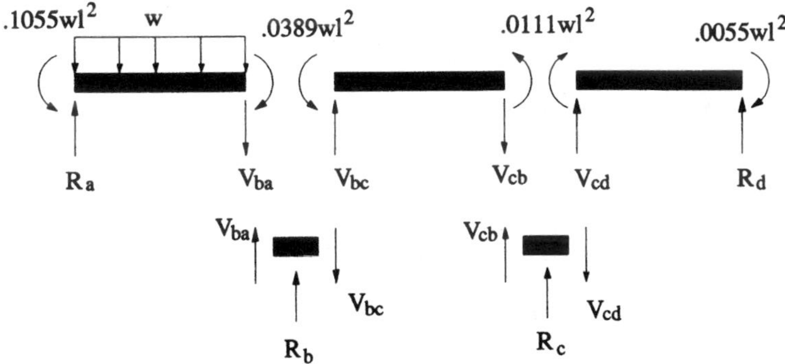

Figure APPC-4

SELECTED REFERENCES

Fleming, J. F., *Computer Analysis of Structural Systems*, McGraw-Hill, New York, 1983.

Grandin, H., *Fundamentals of the Finite Element Method*, Macmillan, New York, 1986.

Livesley, R. K., *Matrix Methods of Structural Analysis*, Pergamon Press, New York, Macmillan, 1964.

Logan, D. L., *A First Course in the Finite Element Method*, PWS Engineering, Boston, 1986.

McGuire, W., and Gallagher, R. H., *Matrix Structural Analysis*, John Wiley, New York, 1979.

Meek, J. L., *Matrix Structural Analysis*, McGraw-Hill, New York, 1971.

Meyers, V. J., *Matrix Analysis of Structures*, Harper and Row, New York, 1983.

Norris, C. H., Wilbur, J. B., and Utku, S., *Elementary Structural Analysis*, McGraw-Hill, New York, 1976.

Roark, R. J., and Young, W. C., *Formulas for Stress and Strain*, McGraw-Hill, New York, 1975.

Rubinstein, M., *Matrix Computer Analysis of Structures*, Prentice Hall, Englewood Cliffs, N.J., 1966.

Sack, R. L., *Matrix Structural Analysis*, PWS Kent, Boston, 1989.

Wang, C., *Intermediate Structural Analysis*, McGraw-Hill, New York, 1983.

Wang, C., *Matrix Methods of Structural Analysis*, International, Scranton, Pa., 1970.

Wang, C., *Structural Analysis on Microcomputers*, Macmillan, New York, 1986.

Weaver, W., and Gere, J. M., *Matrix Analysis of Framed Structures*, Van Nostrand Reinhold, New York, 1990.

ANSWERS TO SELECTED PROBLEMS

Chapter 1

1.1. (a), (b)

$$[K] = \begin{bmatrix} EA/20 & -EA/20 & 0 & 0 \\ -EA/20 & EA/15 + EA/20 & -EA/15 & 0 \\ 0 & -EA/15 & EA/15 + EA/10 & -EA/10 \\ 0 & 0 & -EA/10 & EA/10 \end{bmatrix}$$

(c) $u_1 = 0,\ u_2 = 4.48 \times 10^{-3}\text{in},\ u_3 = 5.26 \times 10^{-3}\text{in},\ u_4 = 6.64 \times 10^{-3}\text{in}.$

1.3. $u_1 = 0$, $u_2 = .01333$ in, $u_3 = .01521$ in.

Member 1: $P_1 = -50{,}000\#$, $P_2 = 30{,}000\#$ **Member 2:** $P_2 = -30{,}000\#$, $P_3 = -15{,}000\#$

1.5. (a) $R_1 = Fb/L,\ R_2 = Fa/L$ (b) $R_1 = kL^2/6,\ R_2 = kL^2/3$

(c) $R_1 = R_2 = L/\pi$ (d) $R_1 = kL^3/12,\ R_2 = kL^3/4$

(e) $R_1 = kL^4/20,\ R_2 = kL^4/5$ (f) $R_1 = w_1L/3 + w_2L/6,\ R_2 = w_1L/6 + w_2L/3$

1.9. $u_{exact} = (2000/60 \times 10^6)[20x - x^2/2]$

4-element model: $u_1 = 0$, $u_2 = .002916''$, $u_3 = .005''$, $u_4 = .00625''$, $u_5 = .00666''$

1.11. $u_2 = .00187''$, $u_3 = -.00089''$, $F_1 = F_2 = 8000\#$

1.13. **(a)** $S = 1,\ K = 8$
(b) $S = 0,\ K = 4$
(c) $S = 3,\ K = 0$ (neglect axial)
(d) $S = 7,\ K = 11$

Chapter 2

2.1. (a)

$$[K] = 29 \times 10^4 \begin{bmatrix} .8333 & 0 & -.8333 & 0 & 0 & 0 \\ 0 & 0 & 0 & 0 & 0 & 0 \\ -.8333 & 0 & 2.0273 & .5970 & -1.1940 & -.5970 \\ 0 & 0 & .5970 & .2985 & -.5970 & -.2985 \\ 0 & 0 & -1.1940 & -.5970 & 1.1940 & .5970 \\ 0 & 0 & -.5970 & -.2985 & .5970 & .2985 \end{bmatrix}$$

(b) $[K]_R = 29 \times 10^4 \begin{bmatrix} 2.0273 & .5970 \\ .5970 & .2985 \end{bmatrix}$

(c) $u_3 = .0828''$, $u_4 = -.2812''$ **Member 1:** $P_3 = 20{,}000\#$ **Member 2:** $P_3 = 22{,}351\#$

2.3. $[K]_R = \frac{29\times 10^6(2)}{120}\begin{bmatrix} .7071 & 0 & 0 \\ 0 & 1.7071 & 0 \\ 0 & 0 & 2 \end{bmatrix}$

$u_3 = .1170''$, $u_4 = -.0364''$, $u_7 = 0$.

Member forces: $P_1 = 19497\#$, $P_2 = 0$, $P_3 = -17574\#$, $P_4 = -37071\#$, $P_5 = 0$.

2.5. **(a)** Reduced equation:

$$\begin{Bmatrix} 5000\# \\ 0 \\ 0 \end{Bmatrix} = \begin{bmatrix} 49690.41 & 0 & -24845.2 \\ 0 & 337650.5 & -49690.41 \\ -24845.2 & -49690.41 & 24845.2 \end{bmatrix} \begin{Bmatrix} u_1 \\ u_2 \\ .375 \end{Bmatrix}$$

$u_1 = .2881''$, $u_2 = .0552''$, $u_3 = .375''$

(b)

$$K_{PP} = \begin{bmatrix} 49690.41 & 0 \\ 0 & 337650.5 \end{bmatrix} \qquad u_P = \begin{Bmatrix} u_1 \\ u_2 \end{Bmatrix}$$

$$\{F_P\} = \begin{Bmatrix} 5000 \\ 0 \end{Bmatrix}$$

Member forces: $P_1 = 9783\#$, $P_2 = -7666.6\#$, $P_3 = -1304.4\#$

2.7. Equivalent nodal load at node 3 = 71201.7#

$u_3 = -.1564''$, $u_4 = -.0408''$, $u_5 = -.1972''$, $u_6 = -.0408''$

Member 1: $P = -16433\#$ **Member 5:** $P = 23268\#$

2.9. $u_1 = u_2 = u_{10} = 0$, $u_3 = .1170''$, $u_4 = .0621''$, $u_5 = .1791''$, $u_6 = -.0621''$, $u_7 = .3032''$, $u_8 = -.7164''$, $u_9 = -.0621''$
Member forces: $P_1 = P_2 = -P_3 = 10000\#$, $P_4 = -P_5 = 14142.1\#$, $P_7 = 20000\#$, $P_8 = -28284.3\#$

2.11. $u_1 = u_2 = u_6 = 0$, $u_3 = -.6114''$, $u_4 = -.0368''$, $u_5 = -.0663''$, $u_7 = -.0331''$, $u_8 = -.0029''$
$P_1 = -66214\#$, $P_2 = -5460\#$, $P_3 = -16020\#$, $P_4 = -6552\#$, $P_5 = 45589\#$, $P_6 = -5460\#$

2.13. Components of equivalent nodal forces are 10115.97 (x) and 20231.94 (y)

$u_1 = u_2 = u_6 = 0$, $u_3 = -.0063''$, $u_4 = .0990''$, $u_5 = -.0126''$, $u_7 = -.0063''$, $u_8 = .0611''$

$P_1 = P_5 = -4096\#$, $P_2 = P_6 = 6107\#$, $P_3 = -3053\#$, $P_4 = 7328\#$

2.15. $u_3 = u_5 = .3224''$, $u_4 = .0368''$, $u_7 = .1939''$
$P_1 = -P_9 = 13333\#$, $P_2 = P_7 = -P_3 = -P_8 = -42164\#$

Chapter 3

3.1.

$$[K]_R = \begin{bmatrix} 232000 & -3867 & 77333 \\ -3867 & 128.89 & -3867 \\ 77333 & -3867 & 154667 \end{bmatrix} \qquad F_R = \begin{Bmatrix} 0 \\ -10^k \\ 0 \end{Bmatrix}$$

$u_4 = -.00776$ rad., $u_5 = -.7759''$, $u_6 = -.01552$ rad.

Member 1: **Member 2:**

$$\begin{Bmatrix} P_1 \\ P_2 \\ P_3 \\ P_4 \end{Bmatrix} = \begin{Bmatrix} -7.5^k \\ -300''^{-k} \\ 7.5^k \\ -600''^{-k} \end{Bmatrix} \qquad \begin{Bmatrix} P_1 \\ P_2 \\ P_3 \\ P_4 \end{Bmatrix} = \begin{Bmatrix} 10.0^k \\ 600''^{-k} \\ -10.0^k \\ 0 \end{Bmatrix}$$

3.3. $[K]_R = [150155.6]$ Equivalent nodal force $= 990''^{-k}(CCW)$

$u_4 = .006593$ rad.

Member 1: **Member 2:**

$$\begin{Bmatrix} P_1 \\ P_2 \\ P_3 \\ P_4 \end{Bmatrix} = \begin{Bmatrix} 2.06^k \\ 247.5''^{-k} \\ -2.06^k \\ 495.0''^{-k} \end{Bmatrix} \qquad \begin{Bmatrix} P_1 \\ P_2 \\ P_3 \\ P_4 \end{Bmatrix} = \begin{Bmatrix} 2.06^k \\ 495''^{-k} \\ -2.06^k \\ 247.5''^{-k} \end{Bmatrix}$$

Add fixed end forces.

3.5.

$$\begin{Bmatrix} -212.66''^{-k} \\ 119.63''^{-k} \end{Bmatrix} = \begin{bmatrix} 141778 & 42533 \\ 42533 & 191400 \end{bmatrix} \begin{Bmatrix} u_4 \\ u_6 \end{Bmatrix}$$

$u_4 = -.001808$ rad, $u_6 = .001027$ rad.
Selected forces:

Member 1: $P_1 = -.85^k$, $P_2 = -51.27''^{-k}$, $P_4 = -102.53''^{-k}$

Member 2: $P_1 = -.83^k$, $P_2 = -110.13''^{-k}$, $P_4 = 10.45''^{-k}$

Member 3: $P_1 = 1.71^k$, $P_2 = 109.18''^{-k}$, $P_4 = 54.59''^{-k}$
Add fixed end forces.

3.7.

$$\begin{Bmatrix} 0 \\ -36^k \\ -2160''^{-k} \end{Bmatrix} = \begin{bmatrix} 1613.6 & 0 & 302.1 \\ 0 & 1210.0 & 295.4 \\ 302.1 & 295.4 & 119222 \end{bmatrix} \begin{Bmatrix} u_4 \\ u_5 \\ u_6 \end{Bmatrix}$$

$u_4 = .003382''$, $u_5 = -.025343''$, $u_6 = -.018063$ rad.

Selected forces:

Member 1: $P_1 = 30.62^k$, $P_3 = -435.5''^{-k}$, $P_5 = 5.45^k$

Member 2: $P_2 = -5.38^k$, $P_3 = -1288''^{-k}$, $P_6 = -647.7''^{-k}$
Add fixed end forces to member 2.

3.9.

$$\begin{Bmatrix} -165.88''^{-k} \\ 0 \end{Bmatrix} = \begin{bmatrix} 212667 & 53167 \\ 53167 & 106333 \end{bmatrix} \begin{Bmatrix} u_4 \\ u_6 \end{Bmatrix}$$

$u_4 = -.000891$ rad, $u_6 = .000446$ rad.

Member 1: $P_1 = -1.48^k$, $P_2 = -47.39''^{-k}$, $P_4 = -94.79''^{-k}$

Member 2: $P_1 = -.74^k$, $P_2 = -71.09''^{-k}$, $P_4 = 0$.
Add fixed end forces to member 1.

3.13.

$$\begin{Bmatrix} -106.66''^{-k} \\ -186.67''^{-k} \\ 400''^{-k} \end{Bmatrix} = \begin{bmatrix} 116000 & 58000 & 0 \\ 58000 & 203000 & 43500 \\ 0 & 43500 & 87000 \end{bmatrix} \begin{Bmatrix} u_2 \\ u_4 \\ u_6 \end{Bmatrix}$$

$u_2 = .000175$ rad, $u_4 = -.002189$ rad, $u_6 = .005692$ rad.

Member 1: $P_1 = -1.95^k$, $P_2 = -106.66''^{-k}$, $P_4 = -243.81''^{-k}$

Member 2: $P_1 = 1.90^k$, $P_2 = 57.14''^{-k}$, $P_4 = 400''^{-k}$
Add fixed end forces.

3.15.

$$\begin{Bmatrix} u_4 \\ u_5 \\ u_6 \\ u_7 \\ u_8 \\ u_9 \end{Bmatrix} = \begin{Bmatrix} -.018684'' \\ -.001221'' \\ -.001136\ rad \\ -.018828'' \\ -.002917'' \\ .001761\ rad \end{Bmatrix}$$

Selected member forces:

Member 1: $P_1 = 5.90^k$, $P_3 = -61.99''^{-k}$, $P_6 = -105.92''^{-k}$

Member 2: $P_2 = 1.58^k$, $P_3 = -45.28''^{-k}$, $P_6 = 234.79''^{-k}$

Member 3: $P_1 = 14.1^k$, $P_3 = 118.01''^{-k}$, $P_6 = 49.9''^{-k}$
Add fixed end forces.

3.17. $u_3 = -.026294$ rad, $u_5 = .005137''$, $u_8 = -.005137''$, $u_{12} = -.04913$ rad.

Selected member forces:

Member 1: $P_1 = -12.41^k$, $P_5 = -4.83^k$, $P_6 = 1158.5''^{-k}$

Member 2: $P_1 = 15.17^k$, $P_3 = -1158.5''^{-k}$, $P_6 = -1820.8''^{-k}$

Member 3: $P_3 = 1820.8''^{-k}$, $P_5 = -15.17^k$

3.19. $u_4 = .000344''$, $u_6 = -.001382$ rad, $u_8 = -.004138''$

Selected member forces:

Member 1: $P_2 = -1.66^k$, $P_3 = -66.36''^{-k}$, $P_6 = -133.14''^{-k}$

Member 2: $P_3 = -66.78''^{-k}$, $P_6 = 66.78''^{-k}$

Member 3: $P_1 = 10^k$, $P_3 = 133.14''^{-k}$
Add fixed end forces to member 2.

Chapter 4

4.1.

$$\begin{Bmatrix} 0 \\ 0 \\ -10^k \end{Bmatrix} = \begin{bmatrix} 132299 & 0 & 0 \\ 0 & 235126.5 & 2448.9 \\ 0 & 2448.9 & 34.01 \end{bmatrix} \begin{Bmatrix} u_1 \\ u_2 \\ u_3 \end{Bmatrix}$$

$u_1 = 0$, $u_2 = .012224$ rad, $u_3 = -1.1755$ in.
Member forces:

Member 1: $P_1 = -.34''^{-k}$, $P_2 = .25''^{-k}$, $P_3 = -5^k$, $P_5 = 899.75''^{-k}$

Member 2: $P_1 = .34''^{-k}$, $P_3 = -5^k$, $P_5 = 899.75''^{-k}$

4.5. Selected displacements:

$u_1 = -.000116$, $u_5 = .002728$, $u_{10} = .001544$, $u_{17} = -.001364$
Selected member forces:

Member 2: $P_1 = 128.28''^{-k}$, $P_4 = -128.28''^{-k}$, $P_6 = -10^k$

Member 5: $P_1 = -20.49''^{-k}$, $P_2 = -720''^{-k}$, $P_6 = -3^k$

Member 6: $P_2 = 107.79''^{-k}$, $P_5 = -107.79''^{-k}$

4.7. Selected displacements:

$u_2 = .00163$ rad, $u_{11} = .001406$ rad, $u_{18} = -.13283''$, $u_{26} = -.00163$ rad.
Selected member forces:

Member 5: $P_1 = 7.69'' - k$, $P_5 = -217.31''^{-k}$, $P_6 = -1.87^k$

Member 9: $P_2 = 15.38''^{-k}$, $P_3 = 1.25^k$, $P_5 = -165.4''^{-k}$

Member 11: $P_2 = -7.69''^{-k}$, $P_3 = 1.87^k$, $P_5 = -217.3''^{-k}$

4.9. Selected displacements:

$u_4 = -.001136$ rad, $u_6 = -.03423''$, $u_{15} = -.12524''$
Selected member forces:

Member 2: $P_2 = 112.5''^{-k}$, $P_3 = -1.88^K$, $P_4 = .15''^{-k}$

Member 9: $P_3 = 1.25^k$, $P_5 = -149.7''^{-k}$, $P_6 = -1.25^k$

Chapter 5

5.1.

$$\begin{Bmatrix} 20 \\ 0 \\ 0 \end{Bmatrix} = \begin{bmatrix} 21.21 & 0 & 0 \\ 0 & 12.13 & -11.00 \\ 0 & -11.00 & 369.76 \end{bmatrix} \begin{Bmatrix} u_1 \\ u_2 \\ u_3 \end{Bmatrix}$$

$u_1 = .94267''$, $u_2 = u_3 = 0$
Axial forces: $P_1 = P_3 = 0$, $P_2 = -P_4 = 22.5^k$

5.3. Selected displacements:
$u_7 = .00274''$, $u_9 = -.00943''$, $u_{12} = -.00943''$
Member forces:
$P_1 = P_2 = 2.207^k$, $P_3 = P_4 = 1.122^k$, $P_5 = 1.023^k$, $P_6 = P_7 = -5.918^k$

5.7. Selected displacements:
$u_{13} = .01166''$, $u_{20} = .00528''$, $u_{24} = -.1347''$, $u_{34} = -.0410''$
Selected member forces:

$P_5 = -9.89^k$, $P_7 = 7.67^k$, $P_{19} = -12.6^k$, $P_{27} = 8.91^k$

Chapter 6

6.1. Selected displacements:
$u_1 = -.00164''$, $u_3 = -.00123$ rad, $u_6 = .0122$ rad.
Selected member forces:

Member 1: $P_1 = 2.07^k$, $P_4 = -44.56^k$, $P_{12} = 99.22^{''-k}$

Member 3: $P_1 = 2.76^k$, $P_6 = 265.6^{''-k}$, $P_{10} = -33.43^k$

6.5. Selected displacements:
$u_7 = -.00387''$, $u_{12} = .00237$ rad, $u_{20} = -.0193''$
Selected member forces:

Member 2: $P_1 = 202.86^{\#}$, $P_6 = 326053^{''-\#}$, $P_{10} = 0$

Member 4: $P_1 = 25208\#$, $P_8 = 3028\#$, $P_{11} = -9737\#$

6.7. Selected displacements:
$u_4 = -.0046$ rad., $u_9 = -.0053''$, $u_{17} = -.0015$ rad.
Selected member forces:

Member 1 : $P_1 = 14.35^k, P_5 = 48.63^{''k}, P_{12} = -80^{''-k}$

Member 4 : $P_1 = 10.61^k, P_6 = 37.02^{''-k}, P_{10} = 65.72^{''k}$

Chapter 7

7.1. maxdiff $= 3$ $bw = 8$

7.3. maxdiff $= 3$ $bw = 8$

7.5. maxdiff $= 2$ $bw = 9$

7.7. maxdiff $= 2$ $bw = 9$

7.9. As numbered, maxdiff $= 4$ $bw = 15$
Exchange 1 and 4, maxdiff $= 3$ $bw = 12$
Renumber 1 to 3, 3 to 4, 4 to 1, maxdiff $= 2$ $bw = 9$

7.11. As numbered, maxdiff $= 4$ $bw = 15$

7.13. As numbered, maxdiff $= 4$ $bw = 30$
Exchange 1 and 3, maxdiff $= 2$ $bw = 18$

7.15. As numbered, maxdiff $= 3$ $bw = 24$

7.17. For the truss members, 4, 5, 7, and 8 will contribute to K_R.

$$[K]_R = \begin{bmatrix} 592.94 & 85.44 & 1450 & -483.33 & 0 & 0 \\ - & 592.94 & 1450 & 0 & -24.17 & 1450 \\ - & - & 232000 & 0 & -1450 & 58000 \\ & sym. & & 592.94 & -85.44 & 1450 \\ - & - & - & - & 592.94 & -1450 \\ - & - & - & - & - & 232000 \end{bmatrix}$$

7.19.

$$[\beta'] = \begin{bmatrix} 1 & 0 & 0 & 0 \\ 0 & 1 & 0 & 0 \\ 0 & 0 & .866 & -.5 \\ 0 & 0 & .5 & .866 \end{bmatrix}$$

$$[\beta']^T[K][\beta'] = \begin{bmatrix} 185611.2 & -123740.8 & -98868.9 & 199965.1 \\ - & 565827.3 & 65912.6 & -133310.1 \\ - & sym. & 294316.6 & -246036.9 \\ - & - & - & 295984.6 \end{bmatrix}$$

Remove row and column 4. With $\{F\} = \{5\ 0\ 0\}$,
$\{u_3 \quad u_4 \quad d_5\} = \{.0374 \quad .00690 \quad .01102\}$
Member forces: 1: 3.33^k 2: 3.06^k 3: -6.01^k

7.21. **(a)**

$$[K]_R = \begin{bmatrix} 2436.80 & 0 & 1208.33 \\ 0 & 2421.69 & 0 \\ 1208.33 & 0 & 96666.66 \end{bmatrix}$$

(b)

$$[K]_R = \begin{bmatrix} 2421.69 & 0 & 0 \\ 0 & 2436.80 & 1208.33 \\ 0 & 1208.33 & 96666.66 \end{bmatrix}$$

7.23. Sub 1 consists of left 4 members; sub 2 as right 4 members.
For sub 1, relabel nodes: 2 to 4, 1 to 3, 4 to 1, 3 to 2.
For sub 2, relabel nodes: 4 to 1, 3 to 2, 6 to 3, 5 to 4.
For sub 1:

$$[K]_R = [K]_c = \begin{bmatrix} 654.22 & 170.88 & 0 & 0 \\ - & 654.22 & 0 & -483.33 \\ - & sym. & 483.33 & 0 \\ - & - & - & 483.33 \end{bmatrix}$$

For sub 2:

$$[K]_c = \begin{bmatrix} 0 & 0 & 0 & 0 \\ 0 & 0 & 0 & 0 \\ 0 & 0 & 0 & 0 \\ 0 & 0 & 0 & 0 \end{bmatrix}$$

$\{F\}_c = \{10 \quad -10 \quad -10 \quad 0\}$
$\{u_1 \quad u_2 \quad u_3 \quad u_4\} = \{.0414 \quad -.0999 \quad -.0207 \quad -.0999\}$ in
$\{u_s\} = \{.0414 \quad -.2412 \quad -.0414 \quad -.2412\}$ in

7.25. **(a)** $u_2 = 9.2 \times 10^{-7} P$
(b) $u_2 = 3.4 \times 10^{-7} P$, $u_3 = 9.2 \times 10^{-7} P$

Chapter 8

8.1. **(a)** $N_1 = 1 - (\sin x/\sin L) \quad N_2 = \sin x/\sin L$
(b) $N_1 = 1 - \sin(\pi x/2L) \quad N_2 = \sin(\pi x/2L)$

8.3. Use equations (8.37) and (8.55).

8.9. $F_{1e} = M\ N_1(a) = M(2a^3/L^3 - 3a^2/L^2 + 1)$

$F_{4e} = M\ N_4(a) = M(a^3/L^2 - a^2/L)$

8.11. $u(x) = -(5F/E)\ln(1 - x/10)$

8.13. $y(L/2) = PL^3/(48.7EI)$

Chapter 9

9.3. $u_5 = .1064$ in, $u_6 = 0$ $\sigma_x = 0$, $\sigma_y = 0$, $\tau_{xy} = 8000$ psi.

9.5. $\{u_3 \quad u_4 \quad u_5 \quad u_6\} = \{.01593 \quad .00250 \quad .01463 \quad .00010\}$ in
Element 1: $\sigma_x = 39808$ psi.
Element 2: $\sigma_x = 40187$ psi.

INDEX